THE MATH TUTOR'S FRIEND

probability & statistics

algebra

trigonometry

pre-algebra

geometry

calculus

BOOK 9

SERIES GUIDE

plus
FLASHMATH
Handouts &
References

by
Joe Cyr

Cyr, Joseph
The Math Tutor's Friend / Book 9 – Series Guide / by Joe Cyr
Third Edition ISBN: 978-1-967039-50-0 Joseph Cyr
Second Edition ISBN: 979-8-9879759-9-2 Joseph Cyr
First Edition ISBN: 9798490274001 Amazon Kindle, 2022

3 4 5 6 7 8 9

Subjects: Main category of book – *tutoring high-school math - pre-algebra; algebra; geometry; trigonometry; probability; statistica; pre-calculus, introduction to calculus*

2.Other category - math tutor reference

Cover illustrations by Ramon E. Owen. Cover Design by Joe Cyr

Book design by Joe Cyr

The Math Tutor's Friend Series

Books

1 - Pre-algebra

2 - Algebra I

3 - Algebra II

4 - Algebra III

5 - Probability & Statistics

6 - Geometry

7 - Trigonometry

8 - Pre-calculus

9 - Series index & "FLASHMATH" handouts/reference cards

THE MATH TUTOR'S FRIEND
TABLE OF CONTENTS
BOOK 9
SERIES GUIDE
(with FLASHMATH handouts/references)

being

handy quick reference notes covering Pre-algebra to Calculus in a series of eight volumes

Sections

•

About the "Math Tutor's Friend" Series:
Teachers impart knowledge guided by lesson plans in a course syllabus. Math tutors, on the other hand, face varying topics at differing levels in successive sessions; for example, pre-algebra, advanced algebra, geometry, trigonometry, statistics, or even calculus! This series addresses the issue and is intended to provide the tutor with needed quick review of high-school math topics. The material is not presented rigorously nor does it abound with proofs and explanation that are available in textbooks - the tutor already knows this stuff - but a quick refresher is always a good move before diving into a tutoring session.
Too many in high-school have reached the 11th and 12th grades lacking basic math skills such as knowing the multiplication table (relying on a calculator to multiply 6 by 7, for example), handling signed numbers, and performing operations involving fractions - knowledge that should have been acquired by the fifth grade! I try to address this deficiency by devoting the first few tutoring sessions to these fundamentals and providing handouts of multiplication tables,etc.

Books 1 through 8 of this series include a number of quiz questions (answers explained) that the math tutor may find useful.
Book 9 is an overall index of topics. It contains a set of summary "FLASHMATH" pages for quick reference or for use as handouts.
This third edition corrects typos found in previous editions and expands discussion of some of the topics.
Happy Tutoring!

Joe Cyr
Sarasota, FL

●

1. CONTENTS OF BOOKS 1 THROUGH 8
Each book in the series contains an index of terms and a section of suggested quiz questions with answers explained.

BOOK 1 - *PRE-ALGEBRA*

Sections

BOOK 2 - *ALGEBRA I*

Sections

BOOK 3 - *ALGEBRA II*

Sections

1. **Absolute Values**
2. **Inequalities**
3. **Factoring & Expanding**
4. **Quadratics**
5. **Rational Functions**
6. **Partial Fraction Expansion**

BOOK 4 - *ALGEBRA III*

Sections

1. **Primes & Number Bases**
2. **Conic Sections**
3. **Complex Number**
4. **Dimension Analysis**
5. **Word Problems**
6. **Interesting Math Concepts**

BOOK 5 - *PROBABILITY & STATISTICS*

Sections

1. **Counting & Probability**
2. **Basic Statistics**
3. **Organizing & Presenting Data**

BOOK 6 - *GEOMETRY*

Sections

1. **Language of Geometry**
2. **Triangles**
3. **Circles**
4. **Geometry Postulates & Theorems**
5. **Logic & Truth Tables**

BOOK 7 - *TRIGONOMETRY*

Sections

1. **The Unit Circle**
2. **Trig Functions**
3. **Polar Coordinates & Vectors**

BOOK 8 - *A LOOK AT CALCULUS*

Sections

1. Limits and Derivatives
2 Integrals
3. Laplace Transforms

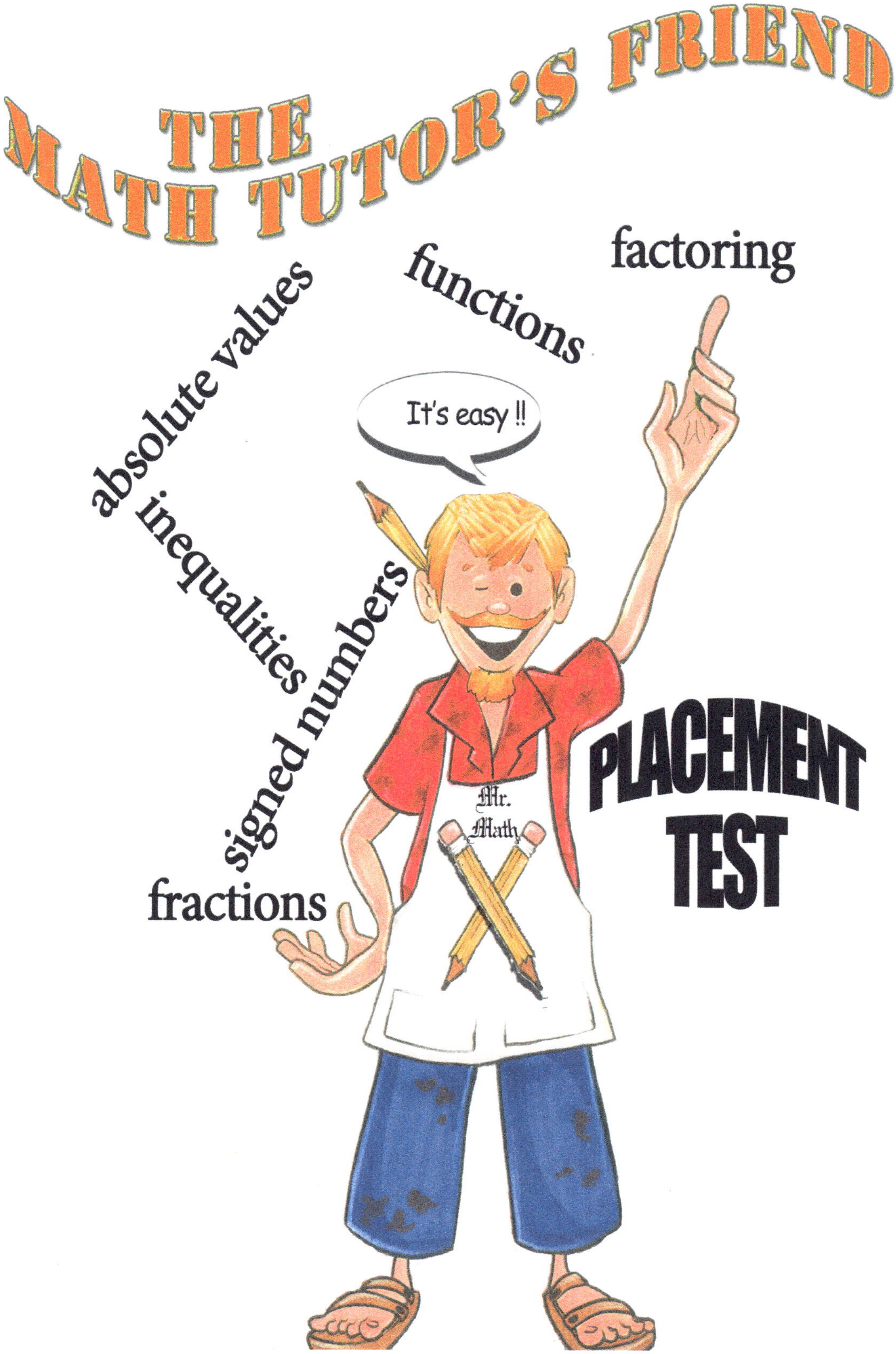
THE
MATH TUTOR'S FRIEND
factoring
functions
absolute values
It's easy !!
inequalities
signed numbers
fractions
Mr.
Math
PLACEMENT
TEST

Administering an informal math placement test may help determine the student's level of knowledge and provide a starting point for tutoring.

PRE-ALGEBRA/ALGEBRA I PLACEMENT EVALUATION

NOTE TO THE STUDENT: If you are not sure of the answer please do not guess; this is not a test for scoring; it is to determine topics for us to work on. ☺ Thanks.

• Cut a cake into eight pieces of equal size (is that possible?) You eat three of the pieces. What fraction of the cake have you eaten?

A. 3/8
B. 8/3
C. 8/8
D. 3/3
E. None of the above.

• Express a comparison of the numbers 23 and -5 as an inequality.

A. $23 > -5$
B. $23 \geq -5$
C. $-5 \leq 23$
D. All of the above
E. None of the above

• Round the decimal 0.5555 to two decimal places

A. 0.56
B. 0.54
C. 0.55
D. 0.556
E. None of the above

• Show the estimated locations of -3 , 0 , 0.245 , ¾ , 1½ , 2 on the number line (you may have to interpolate, i.e., estimate the location).

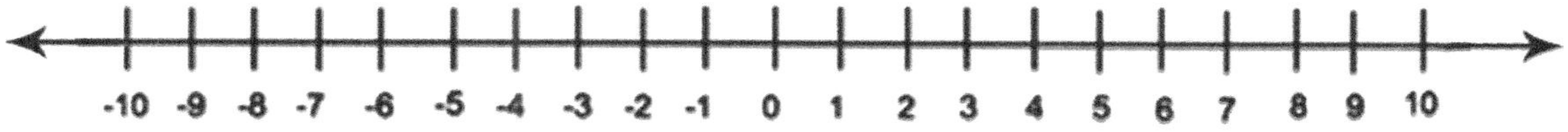

• $19 + 12 + (-3) + 5 + (-10) =$

A. -26
B. 36
C. 26
D. -36
E. None of the above

• Divide 12 by 5/6 (*i.e.*, 12/(5/6) = ?)

A. 12/6
B. 5/72
C. 10
D. 72/5
E. None of the above

• The additive inverse and multiplicative inverse of x^2 are, respectively

A. $-x^2$ and $-1/x^2$
B. $-1/x^2$ and $1/x^2$
C. x and $-x$
D. both $-1/x^2$
E. None of the above

• The multiplicative inverse of 0.5 is

A. -0.5
B. 1/2
C. 2
D. 0.5/1
E. None of the above

• The product of $-987 \times 45 \times (-966) \times (-33) \times 789 \times (-57)$

A. is a transcendental number
B. is a positive number
C. is less than one
D. is an imaginary number
E. None of the above

• 6/37 – 1/8 =

A. 11/296
B. 7/44
C. 5/37
D. 7/8
E. None of the above

• The mixed number $7\frac{1}{2}$ is equivalent to

A. $\frac{15}{2}$
B. $7 + \frac{1}{2}$
C. $\frac{21}{3} + \frac{1}{2}$
D. All of the above
E. None of the above

• $|-34| = ?$

A. -34
B. -3.4
C. 3.4
D. 34
E. None of the above

• You solve a "real-world" math problem and your answer, before simplification, is

$$\frac{1 \text{ mile} \times 60 \text{ seconds} \times 2 \text{ minutes} \times 300 \text{ feet}}{120 \text{ seconds} \times \text{minute} \times 5280 \text{ feet} \times \text{hour}}$$

When simplified, the dimensions of the answer are

A. minutes/feet
B. seconds/hour
C. miles/hour
D. feet/second
E. None of the above

• Express the integer 7 in scientific notation

A. 77
B. 7×10^{0}
C. 7×10^{1}
D. 7×10^{-1}
E. None of the above

• Find the least common multiple (LCM) of 230 and 125

A. 5
B. 230
C. 125
D. 5,750
E. 28,750

- The prime factors (in the complete factorization) of 1,248 are

    A. 25, 3, 13
    B. 2, 4, 8
    C. 3, 7, 19
    D. 1, 2
    E. None of the above

- Rationalize 1/((2-√5))
    A. 2
    B. 5
    C. – (2 +√5)
    D. 2 – √5
    E. None of the above

- (2/3) ÷ (3/4) =

    A. 1/3
    B. 4/9
    C. 4/3
    D. 3/2
    E. None of the above

- The reciprocal of 1.23 is

    A. 1/3.21
    B. 123/100
    C. 123
    D. 100/123
    E. None of the above

- Mathematical operators
    A. Include the infinity symbol (∞)
    B. Separate terms in a mathematical expression
    C. Are signed numbers
    D. Are prime numbers
    E. None of the above

• The expression x^2 signifies

A. The square of x
B. The square root of x
C. That it is a variable different from other subscripted variables
D. The product of x and 2
E. None of the above

• In the proportion expression: $2 : x = 3 : 18$, the variable x is

A. Equal to 6
B. Equal to 18
C. Equal to 12
D Undefined
E. None of the above

• The infinite recurring decimal 1.666... can be written as

A. 5/3
B. 1 2/3
C. 1.667
D. Any of the above
E. None of the above

• A transcendental number can be displayed on the number line.

A. TRUE (why?)
B. FALSE (why not?)

• The value of (-1) – 17 is

A. -27
B. 27
C. -1
D. 1
E. None of the above

• The x-intercept of the line $y = \frac{1}{2} x + 1$ is

A. 1
B. ½
C. -2
D. 0
E. None of the above

• How many terms are there in the following expression?
$y - 11 = 8x^4 - 2x^2 + 8/[(x-2)(x+4)]$

A. 2
B. 7
C. 4
D. 5
E. None of the above

• Solve the inequality for y: $-2y + 3 \leq 2x - 3$

A. $y \geq x$
B. $y \leq -4x + 6$
C. $y \geq x - 6$
D. $y \geq -x + 3$
E. None of the above

• The distance between the points (2, 8) and (5, 4) is

A. 3
B. 4
C. 5
D. 6
E. 25

• Convert 2,350% to a decimal and 35% to a fraction

A. 23.5 and 0.35
B. 2,350 and 35/100
C. 23.5 and 7/20
D. 2.35 and 35/100
E. None of the above

• Assume a 7% sales tax. What is the price, P, of a taxable item sold for $100.00, tax included?

A. $100.07
B. $107.00
C. $93.00
D. $93.46
E. None of the above

• The function $y = 3x^5 - 2x^2 + 10$ has

A. seven solutions
B. five solutions, at least one of which is real
C. two real solutions
D. three solutions, both of which are real
E. None of the above

• The line $y = 2x - 5$ is

A. perpendicular to the line $y = -2x - 5$
B. parallel to the line $-2x - 5$
C. perpendicular to the line $y = -\frac{1}{2}x$
D. parallel to the line $x - 5$
E. None of the above

• The two points (2, 7) and (7, 2)

A. Have a separation of 9 units in the x-direction
B. Have a separation of 5 units along the y-axis
C. Have a separation of 2 units in the x-direction
D. Have a separation of 7 units along the y-axis
E. None of the above

• The distance between the point (5, 8) from the horizontal line $y = 0$ is

A. 0
B. 3
C. 9
D. 8
E. None of the above

• The line $y = x - 3$ is parallel to the line

A. $y = 3x + 1$
B. $x = y + 1$
C. $2x + y = 1$
D. $y = 2x + 2$
E. None of the above

• (1) Simplify $f(x) = (x^2 - 36)/(x + 6)$;
(2) what is the domain of $f(x)$?

A. (1) $f(x) = x - 6$; (2) $x \neq -6$
B. (1) $f(x) = x + 6$; (2) $x \neq -6$
C. (1) $f(x) = x - 6$; (2) all x
D. (1) $f(x) = x + 6$; (2) $x \neq 6$
E. None of the above

• The expression $(a^2 - 2)x^2 + 3ax + 4$ is a

A. trinomial
B. polynomial of four terms
C. polynomial of five terms
D. Polynomial of seven terms
E. None of the above

• The expression equivalent to $7x + 3y + 4x - 4y$ is

A. $10xy$
B. $10x$
C. $11x - y$
D. $11x - 8y$
E. None of the above

• Apply the Distributive Law to the expression $-3(2x - 3y)$

A. $(6x - 9y)$
B. $(2x - 3y)$
C. $(5x - 6y)$
D. $(-6x + 9y)$
E. None of the above

• Simplify the expression $a^2x^3 - 2ax^2 + ax$ by factoring

A. $a(a^2x^3 - 2ax^2 + ax)$
B. $a^2x^2(a^2x^3 - 2ax^2 + ax)$
C. $x^2(a^2x^3 - 2ax^2 + ax)$
D. $a^2x(a^2x^3 - 2ax^2 + ax)$
E. None of the above

• A line segment has end-point coordinates at (-5, 12) and (5, 2). The coordinates of the point that bisects the line segment (center of the line segment) are:

A. (0, 7)
B. (2.5, 6)
C. (-2.5, 6)
D. (7, 0)
E. None of the above

• Solve for x $(3x - 5)/4 = 2x + 5$

A. $x = 5$
B. $x = -5$
C. $x = 0$
D. $x = -4$
E. None of the above

• The slope of the line passing through the points (2, 0) and (0, -3) is

A. 1.5
B. 0
C. -1
D. -5
E. None of the above

• Factor the following expression completely $4x^2 - 9y^2$

A. $4x + 9y$
B. $2x^2 - 3y^2$
C. $(x + y)(x - y)$
D. $(2x - 3y)(2x + 3y)$
E. Cannot factor the expression any further

• A student simplifies $3(2x - 1) - (3x + 1)$ as follows:

Step 1: $3(2x - 1) - (3x + 1) = 6x - 3 - (3x + 1)$
Step 2: $6x - 3 - (3x - 1) = 6x - 3 - 3x + 1$
Step 3: $6x - 3 - 3x + 1 = 3x - 3 + 1$
Step 4: $3x - 3 + 1 = 3 - 2$ [Student's Answer]

A. The answer is correct.
B. The answer is incorrect; mistake in Step 1
C. The answer is incorrect; mistake in Step 2
D. The answer is incorrect; mistake in Step 3
E. The answer is incorrect; mistake in Step 4

- Match each example with the relevant property of equality

A. Reflexive	(1) if a = b and b = c then a = c
B. Symmetric	(2) $a = a$
C. Transitive	(3) if a = b then b can replace "a" in any expression
D. Substitution	(4) If a = b then b = a
E. Multiplication	(5) If a = b then a + c = b + c
F. Addition	(6) If a = b then $a \times$ c = b $\times$ c

A___; B___; C___; D___; E___; F___

- The equation "$ax^2 + bx + c = 0$" contains how many terms?
  - A. 2 terms
  - B. 3 terms
  - C. 4 terms
  - D. 6 terms
  - E. None of the above because it is an equation

- The algebraic expression equivalent to (5/2) $(10x^3 - 2/5\ x)$ is
  - A. $50x^2 - 2$
  - B. $10x^2 - 1$
  - C. $x(25x^2 - 1)$
  - D. $x^2(5x - 2)$
  - E. None of the above

- The degrees of the four terms in the expression $y^2 - x^3/y + x^2 - 32$ are, respectively,
  - A. two, two, two, zero
  - B. two, two, two, two
  - C. two, three, two, negative two
  - D. two, one, two, two
  - E. None of the above

- Which is true for x in the inequality $-4x + 3 \geq 2x - 6$?
  - A. $x \geq -3$
  - B. $x = 3$
  - C. $x \leq 3$
  - D. $x \geq 6$
  - E. None of the above

• Line A in the diagram is expressed as $y = x + 5$. Choose the equation that could be a reasonable equation for Line B.

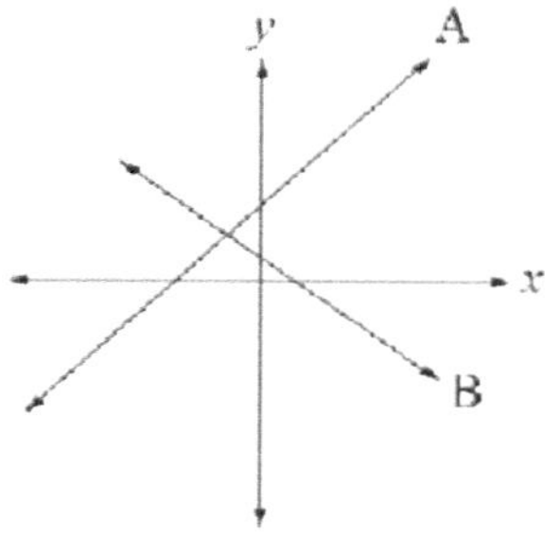

A. $y = 0.6x + 2$
B. $y = -0.6x + 2$
C. $y = x$
D. $y = -x$
E. None of the above

• An expression equivalent to $(x - 2)/y = 5/3x$ is
A. $3x(x - 2) = 5y$
B. $3x^2 - 6x = 5y$
C. $y = (3x^2 - 6x)/5$
D. All are equivalent
E. None are equivalent

• The sum of two integers is 27. The difference is 5.
The numbers are
A. 27 and 5
B. 32 and 27
C. 16 and 11
D. 11 and 6
E. None of the above

• If $x + 12 = 14$, evaluate $x^3 - x - 2 + 6x - 1$
A. 15
B. 12
C. 10.75
D. 11.25
E. None of the above

• The parent function for the function $y = (x - 5)^3 + 13$ is
A. $y = (x - 5)^3$
B, $y = 13$
C. $y = x$
D. $y = x^3$
E. None of the above.

• The graph of a hyperbola shown below (dashed lines are not part of the graph)

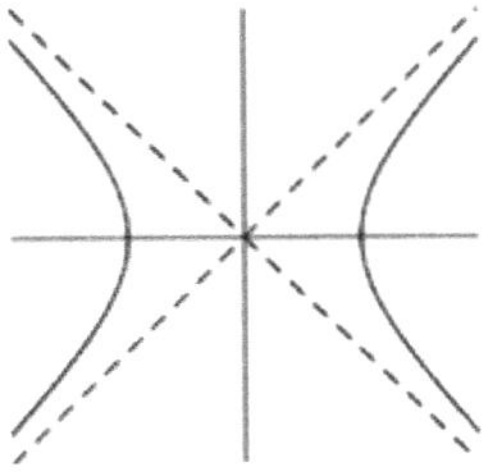

A. is a function?
B. is not a function
C. may not be a function for certain values of x
D. has domain of all x
E. None of the above

• Let $y = g(x)$. Which is the independent variable?
A. x
B. y
C. g
D. All of the above
E. None of the above

• Let $f(x) = x^2 + x - 2$. Then f(3) =
A. 10
B. -1
C. 5
D. 9
E. None of the above

• Let $f(x) = \sqrt{(x-3)}$ (i.e., the square root of $(x - 3)$. The domain of this function is
A. All x such that $x > 0$
B. All x such that $x < 0$
C. The interval $[0, \infty)$
D. The interval $(-\infty, 3)$
E. None of the above

• Three consecutive odd numbers add to 57. What is the middle number?

- In math, the symbols "+" "–" "÷" and "×" are called
    - A. division
    - B. terms
    - C. operators
    - D. exponents
    - E. None of the above

- $[2(32 - 2) + 1] \div 3 + (2 \times 10) - 4 = ?$
    - A. 31
    - B. 20
    - C. 24
    - D. 21
    - E. None of the above

•

THE MATH TUTOR'S FRIEND
math fundamentals
algebra
trigonometry
It's easy !!
geometry
statistics
calculus
Mr. Math
FLASHMATH

by Joe Cyr

2 - FLASHMATH Handouts/References

2-FLASHMATH Handouts/References (cont'd)

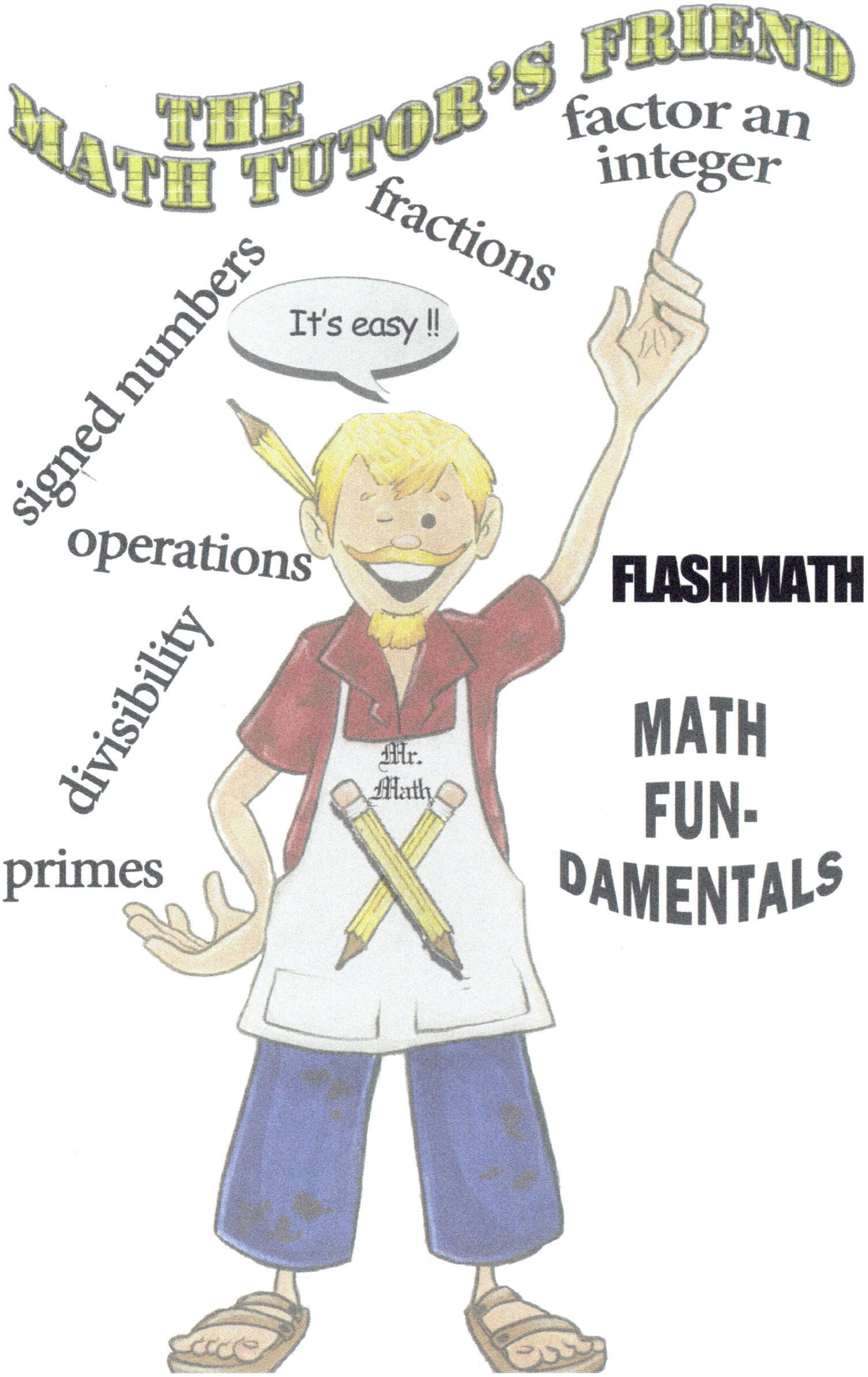
THE MATH TUTOR'S FRIEND
factor an integer
fractions
signed numbers
It's easy !!
operations
FLASHMATH
divisibility
Mr. Math
MATH FUN-DAMENTALS
primes

Anatomy of a Division

FLASHMATH

ANATOMY OF A DIVISION

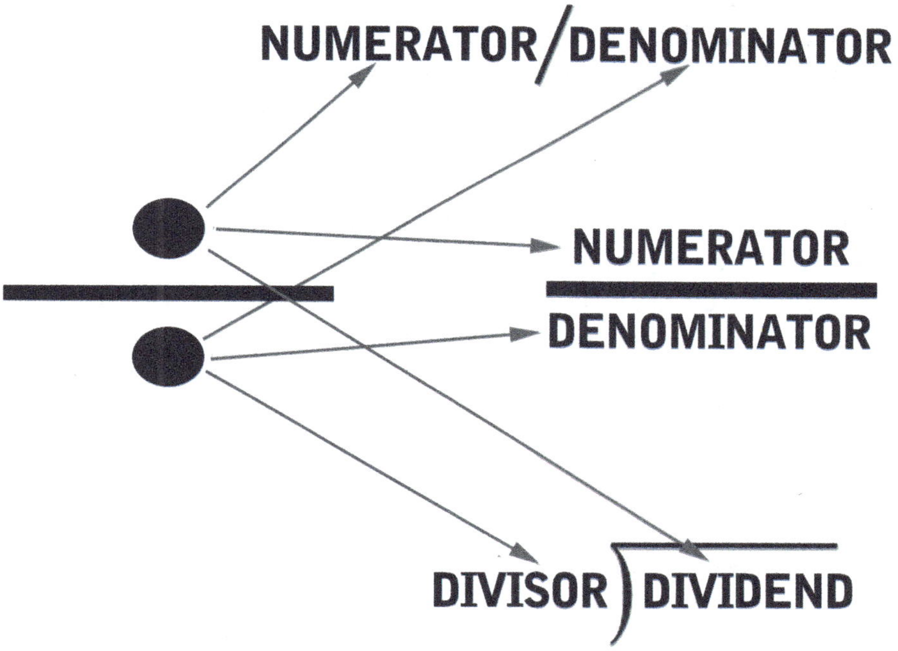

Tips for solving word problems

FLASHMATH

TIPS FOR SOLVING WORD PROBLEMS

- **PUT DOWN YOUR PENCIL AND CALCULATOR**
- **READ THE ENTIRE PROBLEM STATEMENT**
- **WHAT KIND OF PROBLEM IS IT?**
  (*e.g.*, counting, transversal, similar triangles, ...)
- **FIND OUT WHAT'S NEEDED**
- **WHAT ARE YOU ASKED TO FIND?**
- **WHAT DATA ARE PROVIDED?** If the variable is not specified in the problem (*e.g.*, "find the width, w"), give it a variable name: d (for distance, or perhaps depth), t (for time), r (for rate, or perhaps radius), A (for area), *etc.*, or just *x*.
- **DETERMINE THE APPLICABLE FORMULA(S).** Write them down for reference.
- **IS YOUR SOLUTION TO BE EXACT OR APPROXIMATE** (if so, what accuracy is specified)? "$3\pi - \sqrt{2}$" is an exact solution; "8.01" is approximate to two decimal places.
- **SET UP THE PROBLEM.** Read the problem again and write your formulas, equations, or whatever.
- **MAKE NEEDED UNIT CONVERSIONS** You don't want to mix inches and feet, for example.
- **USE DIMENSION ANALYSIS** to set up your solution and to check your answer.
- **DO THE WORK.** Don't rely on the calculator alone; use paper and make sketches; write neatly.
- **CHECK YOUR WORK!** Is the answer reasonable? Is the answer expressed in the specified or proper units (*e.g.*, if the problem data are expressed in feet, then your answer should be in feet or square feet or cubic feet, *etc.*, unless specified otherwise in the problem statement).

Completely factor an integer

FLASHMATH

COMPLETELY FACTOR AN INTEGER

Example: Completely factor 793 800

NOTE: Use divisibility rules for 2, 3, 5, etc...

Divisor	Number
2	793800
2	396900
2	198450
3	99225
3	33075
3	11025
3	3675
5	1225
5	245
7	49
	7

The complete factoring of 793 800:

$$793\,800 = 2 \times 2 \times 2 \times 3 \times 3 \times 3 \times 3 \times 5 \times 5 \times 7 \times 7$$

$$= 2^3 \times 3^4 \times 5^2 \times 7^2$$

Cross multiplying

Add two fractions by cross-multiplying

$$\frac{4}{5} + \frac{2}{3} = \frac{12+10}{15}$$

Multiply the denominators

Divisibility rules

FLASHMATH

DIVISIBILITY RULES

A number is divisible by

2 if the number is even*.

3 if sum of its digits is divisible by 3

4 if the last two digits are divisible by 4

5 if the unit (last) digit is 5 or 0

6 if even* and divisible by 3

7 if 9 times the units digit subtracted from the remaining digits of the number is divisible by 7

8 if its last three digits are divisible by 8

9 if the sum of its digits is divisible by 9

10 if the unit (last) digit is 0

* Decimals are neither odd nor even

Note: There are other rules for divisibility by 11, etc., but those are too involved to be useful here.

Inequalitiers& intervals

SYMBOLOGY

$a = b$	***a*** **is equal to *b***
$a \neq b$	***a*** **is not equal to *b*(also "a < > b")**
$a < b$	***a*** **is less than *b* (strict inequality)**
$a > b$	***a*** **is greater than *b* (strict inequality)**
$a \geq b$	***a*** **is greater than or equal to *b* (weak inequality) <Alt> 242**
$a \leq b$	***a*** **is less than or equal to *b* (weak inequality) <Alt> 243**
$a << b$	***a*** **is much less than *b***
$a >> b$	***a*** **is much greater than *b***

NOTE: *a* cna *b* may be real numbers or algebraic expressions

CLOSED INTERVAL NOTATION

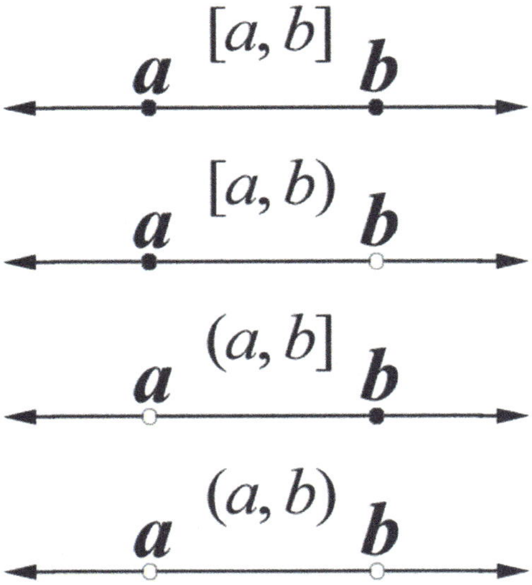

Greek alphabet

Αα ALPHA	Ββ BETA	Γγ GAMMA	Δδ DELTA	Εε EPSILON	Ζζ * ZETA
Ηη* ETA	Θθ THETA	Ιι* IOTA	Κκ* KAPPA	Λλ LAMBDA	Μμ MU
Νν NU	Ξξ * XI	Οο* OMICRON	Ππ PI	Ρρ RHO	Σσ SIGMA
Ττ TAU	Υυ* UPSILON	Φφ PHI	Χχ* CHI	Ψψ * PSI	Ωω OMEGA

*** Not generally used in high-school math**

Handling signed numbers

FLASHMATH

HANDLING SIGNED NUMBERS

HOW TO ADD AND SUBTRACT SIGNED NUMBERS (Algebraic addition)

CASE 1: **Add three or more signed numbers**

(A) - If the numbers all have the same sign (+ or -) just add them and keep the sign

(B) - If there is a mixture of positive and negative numbers

Add all positive numbers to get a positive sum, then add all negative numbers to get a negative sum. We now have just two signed numbers to add. Go to *Case 3* below.

CASE 2: **Subtract two signed numbers (subtract the second number from the first. (the first number is called the *minuend*, and the second number is called the *subtrahend).***

Simply change the sign of the subtrahend and add. Go to *Case 3* below.

CASE 3: **Add two signed numbers**

(A) – If the signs of both numbers are the same (both positive, or both negative) then just add them and attach the sign (positive or negative).

(B) – If the signs are different (that is, one number is positive, the other negative) First, ignore the signs. "Take away" the value of the smaller* number from that of the larger The sign of the answer will be that of the larger number.

HOW TO MULTIPLY AND DIVIDE TWO SIGNED NUMBERS

CASE 1: **If the signs are the same (that is, both positive or both negative), then multiplying or dividing always produces a POSITIVE number.**

CASE 2: **If the signs are different (that is, one number is positive, the other number is negative), then multiplying or dividing always produces a NEGATIVE number.**

NOTE: When multiplying more than two signed numbers, if there is an *odd* number of negative numbers, then the product will be negative; otherwise, the product will be positive.

HOW TO RAISE SIGNED NUMBERS TO A POWER

CASE 1: **Raising a number (base) to a positive power.**

If the base is positive the result is positive.

If the base is negative and the exponent is even, then the result is positive

If the base is negative and the exponent is odd, then the result is negative.

CASE 2: **Raising a number (base) to a negative power.**

Take the reciprocal of the number (base) and the exponent becomes positive. Go to *Case 1.* above.

*** *i.e.*, the absolute value**

Interesting number sequences

FLASHMATH

INTERESTING NUMBER SEQUENCES

9	×	9	+	7	=	88
98	×	9	+	6	=	888
987	×	9	+	5	=	8888
9876	×	9	+	4	=	88888
98765	×	9	+	3	=	888888
987654	×	9	+	2	=	8888888
9876543	×	9	+	1	=	88888888
98765432	×	9	+	0	=	888888888

1	×	9	+	2	=	11
12	×	9	+	3	=	111
123	×	9	+	4	=	1111
1234	×	9	+	5	=	11111
12345	×	9	+	6	=	111111
123456	×	9	+	7	=	1111111
1234567	×	9	+	8	=	11111111
12345678	×	9	+	9	=	111111111
123456789	×	9	+	10	=	1111111111

1	×	8	+	1	=	9
12	×	8	+	2	=	98
123	×	8	+	3	=	987
1234	×	8	+	4	=	9876
12345	×	8	+	5	=	98765
123456	×	8	+	6	=	987654
1234567	×	8	+	7	=	9876543
12345678	×	8	+	8	=	98765432
123456789	×	8	+	9	=	987654321

Math symbols & < Alt > codes

FLASHMATH

MATH SYMBOLS & <Alt> CODES

A TUTORIAL ON MATHEMATICAL SYMBOLS
(with corresponding <Alt> computer keyboard codes)

ab or $a{\cdot}b$ or $a{\times}b$ or $(a)(b)$ "*a* times *b*"
(*i.e.*, the product of *a* and b)

$a = b$ *a* is equal to *b*

$a \neq b$ *a* is not equal to *b* <Alt> 8800

$a < b$ *a* is less than *b* (strict inequality)

$a > b$ *a* is greater than *b* (strict inequality)

$a \geq b$ *a* is greater than or equal to *b*
(weak inequality) <Alt> 242

$a \leq b$ *a* is less than or equal to *b*
(weak inequality) <Alt> 243

$a << b$ *a* is much less than *b* <Alt> 174

$a >> b$ *a* is much greater than *b* <Alt> 175

$a \approx b$ *a* is approximately equal to *b* <Alt> 247

$a \equiv b$ Identity: *a* is identical to *b* <Alt> 240
(this means that if *b* changes, then *a* also changes.
Contrast this with the equation $a = b$,
where if *b* changes, then equality is no longer true.)

$a \pm b$ *a* plus or minus *b* <Alt> 241
(*i.e.*, two solutions: $a + b$ and $a - b$)

a/b a fraction with numerator *a* and denominator *b*;
always means "*a* divided by *b*."
(Try to avoid the " ÷ " symbol)

NOTES:

- ***a* and *b* may be real numbers, constants, expressions, or functions.**
- **Use keypad for <Alt> codes (<Num Lock> must be ON)**
- **<Alt> codes may not work with some apps.**

Multiplication table

FLASHMATH

MULTIPLICATION TABLE

×	1	2	3	4	5	6	7	8	9	10
1	1	2	3	4	5	6	7	8	9	10
2	2	4	6	8	10	12	14	16	18	20
3	3	6	9	12	15	18	21	24	27	30
4	4	8	12	16	20	24	28	32	36	40
5	5	10	15	20	25	30	35	40	45	50
6	6	12	18	24	30	36	42	48	54	60
7	7	14	21	28	35	42	49	56	63	70
8	8	16	24	32	40	48	56	64	72	80
9	9	18	27	36	45	54	63	72	81	90
10	10	20	30	40	50	60	70	80	90	100

Shaded cells contain perfect squares

Nomenclature for the four operations

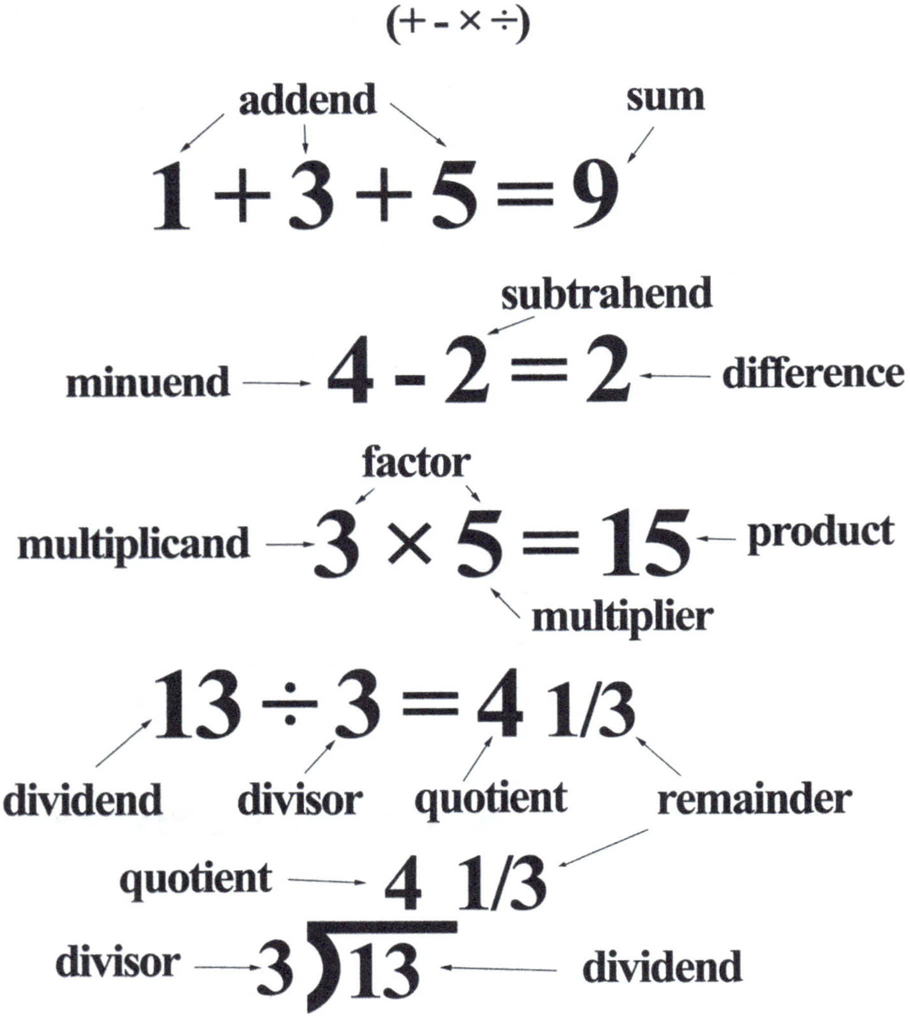

Order of operations

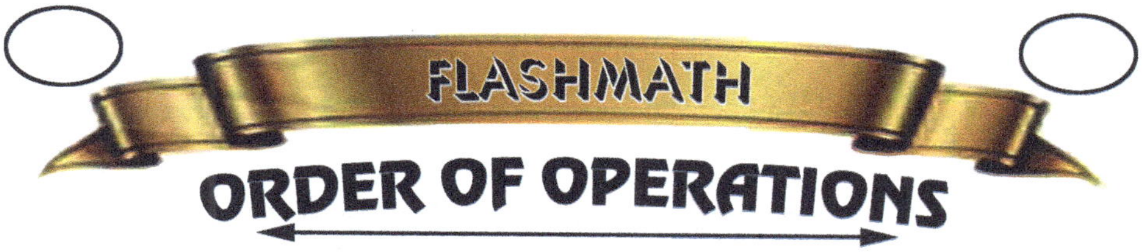

(1) Do all operations inside parentheses; note that the numerators and denominators of fractions serve as parentheses. If parentheses are nested, e.g., {[()+()]-()}, work from the inside out.

(2) Do any work with exponents and radicals.

(3) Working from left to right, do all multiplications and divisions.

(4) Finally, working from left to right, do all additions and subtractions.

NOTE to Rule (4): Recognize that subtracting one number, say 4, from another, say 3, is really only adding a negative number, and the "left to right" rule is unnecessary. Thus,

$$3 - 4 = 3 + (-4) = -1.$$

Another example:

$$26 - 4 - 2 + 5 + 1 - 3 = ?$$

If you work left to right, you would get

$$22 - 2 + 5 + 1 - 3 = 20 + 5 + 1 - 3 = 25 + 1 - 3$$
$$= 26 - 3 = 23.$$

Alternatively, you could group (collect) and sum all of the "pluses" and all of the "minuses" to get

$$(26 + 5 + 1) - (4 + 2 + 3) = 32 - 9 = 23,$$

again, the correct answer (this is the method that I recommend).

FLASHMATH

PRIME NUMBERS

A ***prime*** number is a whole number greater than 1 whose only factors are 1 and itself. A factor is a whole number that can be divided evenly into another number. Numbers that have more than two factors are called ***composite*** numbers. The table below shows the first **300** primes (there are an infinite number of primes). Every composite number can be completely factored into the product of primes. For example, the composite number 793,800 can be completely factored as shown below:

$$2 \times 2 \times 2 \times 3 \times 3 \times 3 \times 3 \times 5 \times 5 \times 7 \times 7 = 793{,}800$$

The First 300 Primes

(the 1,000th prime is 7919)

2	3	5	7	11	13	17	19	23	29
31	37	41	43	47	53	59	61	67	71
73	79	83	89	97	101	103	107	109	113
127	131	137	139	149	151	157	163	167	173
179	181	191	193	197	199	211	223	227	229
233	239	241	251	257	263	269	271	277	281
283	293	307	311	313	317	331	337	347	349
353	359	367	373	379	383	389	397	401	409
419	421	431	433	439	443	449	457	461	463
467	479	487	491	499	503	509	521	523	541
547	557	563	569	571	577	587	593	599	601
607	613	617	619	631	641	643	647	653	659
661	673	677	683	691	701	709	719	727	733
739	743	751	757	761	769	773	787	797	809
811	821	823	827	829	839	853	857	859	863
877	881	883	887	907	911	919	929	937	941
947	953	967	971	977	983	991	997	1009	1013
1019	1021	1031	1033	1039	1049	1051	1061	1063	1069
1087	1091	1093	1097	1103	1109	1117	1123	1129	1151
1153	1163	1171	1181	1187	1193	1201	1213	1217	1223
1229	1231	1237	1249	1259	1277	1279	1283	1289	1291
1297	1301	1303	1307	1319	1321	1327	1361	1367	1373
1381	1399	1409	1423	1427	1429	1433	1439	1447	1451
1453	1459	1471	1481	1483	1487	1489	1493	1499	1511
1523	1531	1543	1549	1553	1559	1567	1571	1579	1583
1597	1601	1607	1609	1613	1619	1621	1627	1637	1657
1663	1667	1669	1693	1697	1699	1709	1721	1723	1733
1741	1747	1753	1759	1777	1783	1787	1789	1801	1811
1823	1831	1847	1861	1867	1871	1873	1877	1879	1889
1901	1907	1913	1931	1933	1949	1951	1973	1979	1987 … •

Note that all primes end in 1, 3, 7. or 9.

Rapid calculation techniques

FLASHMATH

RAPID CALCULATION TECHNIQUES

Mentally Square a 2-digit number ending in 5
Multiply the tens digit by the next higher digit & tack on 25.

Mentally Square a 2-digit number beginning with "5"
Tack the square of the ones digit to 25 plus the ones digit. If the ones digit is 3 or less, precede the square with a place-holder zero; e.g., $52^2 = 2704$; $51^2 = 2601$; but $54^2 = 2916$.
;
Multiply two 2-digit numbers whose last digits sum to 10
Multiply the tens digit by the next higher digit, and "tack on" the product of the units digits

Mentally multiply a number by 5, 25, or 125
5: multiply by 10 and divide by 2; ***25***: multiply by 100 and divide by 4; ***125***: multiply by 1000 and divide by 8.

Rapidly add a column of numbers

(a) Add the 1st column digits assuming other columns are zero (*e.g.*, 3,456 would be 3,000)
(b) Add the 2nd column digits the same way
(c) Continue with all other columns
(d) Answer is the sum of (a), (b), …

Example: Add the following

	3,541
	2,133
	434
	1,267
6,000 (left-most column)	689
+ 1,800 (next column)	21
+ 260 (column)	8,085
+ 25 (right-most column)	
= 6,000 + 1,000 + 800 + 200 + 60 + 25	

Ratio & proportion

FLASHMATH

RATIO & PROPORTION

Proportion:

"c" is to "d" as "a" is to "b"
"a" and "d" are "extremes"
"b" and "c" are "means"

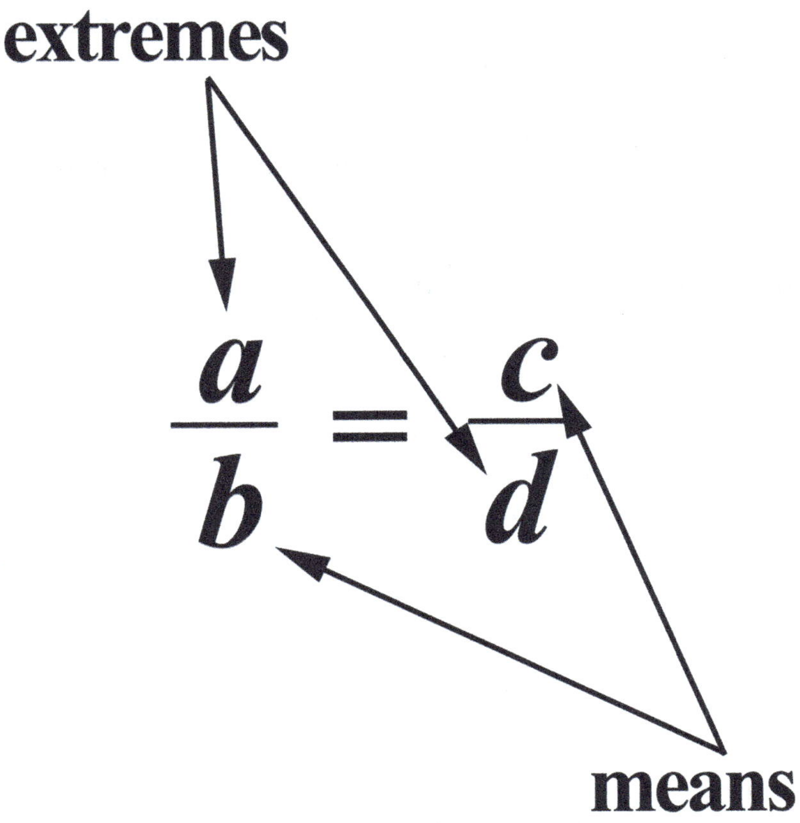

Reading & writing numbers

HOW TO READ AND WRITE NUMBERS

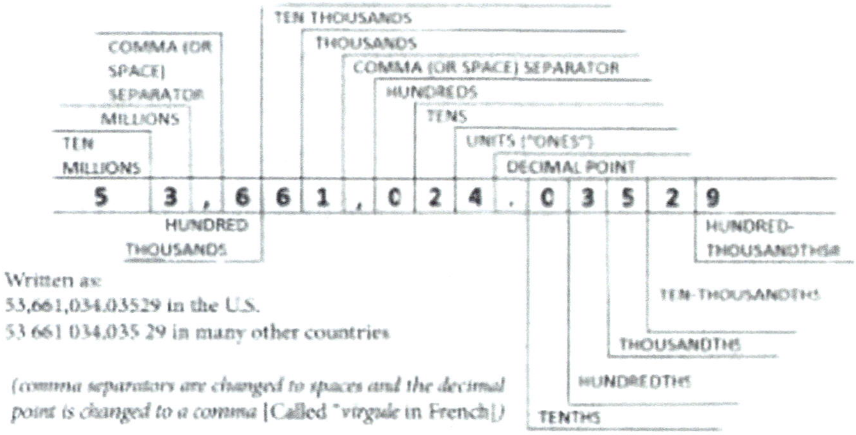

Written or spoken as "fifty-three million six hundred sixty-one thousand twenty-four point zero three five two nine" or "fifty-three million six hundred sixty-one thousand twenty-four and three thousand five hundred twenty-nine hundred-thousandths."

RULES FOR WRITING NUMBERS (in essays, reports, etc.)

Small and large numbers . . .

Spell out whole numbers one through nine. Write larger numbers and decimal numbers in numerals (*e.g.*, 17). Spell out numbers that begin a sentence (consider re-arranging the sentence to avoid the first words being numbers).

Paired numbers . . .

When numbers are paired in a sentence spell out one of them (*e.g.*, "There were two 14-year old girls...").

Hyphenating rules for numbers . . .

Hyphenate two-word numbers between 21 and 99, *e.g.*, thirty-two. Do not hyphenate numbers greater than 99, *e.g.*, three hundred (but three hundred twenty-one).

Repeating decimals

Repeating Decimals

A decimal representation of a number whose decimal digits are periodic (repeating its values at regular intervals) and the infinitely-repeated portion is not zero. For example, the decimal representation of ⅓ is 0.333.... becomes periodic just after the decimal point, repeating the single digit "3" forever. A more complicated example is 3227/555, whose decimal, 5.8144144144..., becomes periodic after the second digit following the decimal point and then repeats the sequence "144" forever. The table below shows two ways to write repeating decimals.

Fraction	Ellipsis	Vinculum
1/9	0.111...	$0.\overline{1}$
1/3	0.333...	$0.\overline{3}$
2/3	0.666...	$0.\overline{6}$
9/11	0.8181...	$0.\overline{81}$
7/12	0.58 333...	$0.58\overline{3}$
1/81	0.012345679...	$0.\overline{012345679}$
22/7	3.142857 142857...	$3.\overline{142857}$

Roman numerals

FLASHMATH

ROMAN NUMERALS

Roman numerals are a system of numerical notations used by the Romans:

1	I	5	V
10	X	50	L
100	C	500	D
1,000	M		

The number 1,782 is MDCCLXXXII in Roman numerals. When required, numbers are denoted by preceding the symbol for 5, 50, 10, 100, etc., with a symbol indicating subtraction. For example, 4 is denoted as IV (5 minus 1), 9 as IX (10 minus 1), 40 as XL (50 minus 10), etc. (an exception is on the faces of clocks, where IIII is sometimes used instead of IV). The following table shows Roman numerals for the first thirty natural numbers.

1	I	6	VI	11	XI	16	XVI	21	XXI	26	XXVI
2	II	7	VII	12	XII	17	XVII	22	XXII	27	XXVII
3	III	8	VIII	13	XIII	18	XVIII	23	XXIII	28	XXVIII
4	IV	9	IX	14	XIV	19	XIX	24	XXIV	29	XXIX
5	V	10	X	15	XV	20	XX	25	XXV	30	XXX

Scientific notation

FLASHMATH

SCIENTIFIC NOTATION

Prefixes for large and small numbers

Prefix	Symbol	1000^m	10^n	Decimal	English word
yotta	Y	1000^8	10^{24}	1000000000000000000000000	septillion
zetta	Z	1000^7	10^{21}	1000000000000000000000	sextillion
exa	E	1000^6	10^{18}	1000000000000000000	quintillion
peta	P	1000^5	10^{15}	1000000000000000	quadrillion
tera	T	1000^4	10^{12}	1000000000000	trillion
giga	G	1000^3	10^9	1000000000	billion
mega	M	1000^2	10^6	1000000	million
kilo	k	1000^1	10^3	1000	thousand
hecto	h	$1000^{2/3}$	10^2	100	hundred
deca	da	$1000^{1/3}$	10^1	10	ten
		1000^0	10^0	1	one
deci	d	$1000^{-1/3}$	10^{-1}	0.1	tenth
centi	c	$1000^{-2/3}$	10^{-2}	0.01	hundredth
milli	m	1000^{-1}	10^{-3}	0.001	thousandth
micro	μ	1000^{-2}	10^{-6}	0.000001	millionth
nano	n	1000^{-3}	10^{-9}	0.000000001	billionth
pico	p	1000^{-4}	10^{-12}	0.000000000001	trillionth
femto	f	1000^{-5}	10^{-15}	0.000000000000001	quadrillionth
atto	a	1000^{-6}	10^{-18}	0.000000000000000001	quintillionth
zepto	z	1000^{-7}	10^{-21}	0.000000000000000000001	sextillionth
yocto	y	1000^{-8}	10^{-24}	0.000000000000000000000001	septillionth

THE
MATH TUTOR'S FRIEND
parent
functions
complex plane
absolute value
It's easy !!
quadratics
asymptotes
conic
sections
Mr.
Math
FLASHMATH
ALGEBRA

Absolute value

FLASHMATH

ABSOLUTE VALUE

The ***absolute value*** of a number is the magnitude of a number without its sign; it is the distance of the number from 0 on the number line and so is always positive or zero. Just think of the absolute value of a number as the number stripped of its sign:

$|3| = 3$ and $|-3| = 3$.

The ***absolute value function*** is a function that contains an algebraic expression within absolute value symbols "| |". The absolute-value parent function is $f(x) = |x|$, and is defined as follows: $|x| = x$ if $x > 0$ and $-x$ if $x < 0$. $|x| = 0$ if $x = 0$.

Properties of absolute values

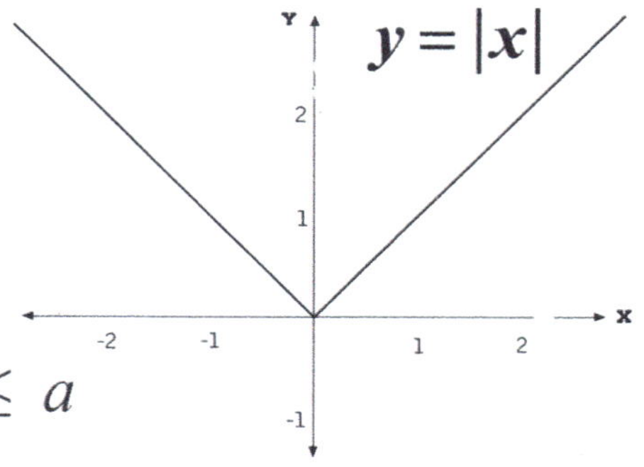

$|x| \geq 0$

$|xy| = |x| \times |y|$

$|x + y| \leq |x| + |y|$

If $|x| \leq a$ then $-a \leq x \leq a$

If $|x| \geq a$ then $x \leq -a$ or $x \geq a$

$|x - y| < |z - w|$ if and only if $(x - y)^2 < (z - w)^2$

$|x| = \sqrt{x^2}$

Absolute value to inequaltiy

ABSOLUTE VALUE TO INEQUALITY CONVERSION

- If $|\boldsymbol{x}| = \boldsymbol{a}$ then $-a < x < a$

- To convert the compound inequality

$$a < x < b$$

to an absolute-value inequality, we write

$$|\,2x - (a + b)\,| < (b - a).$$

Conic sections

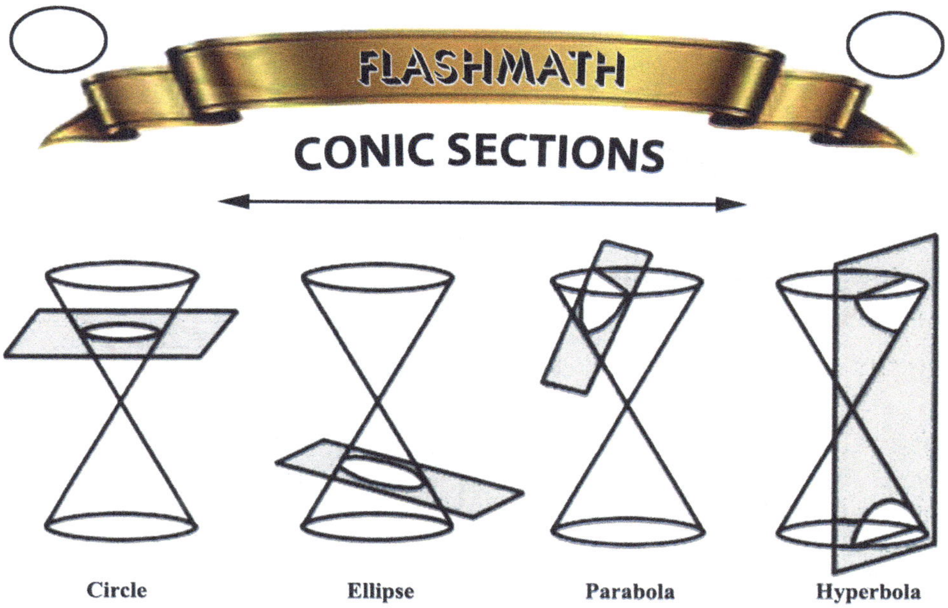

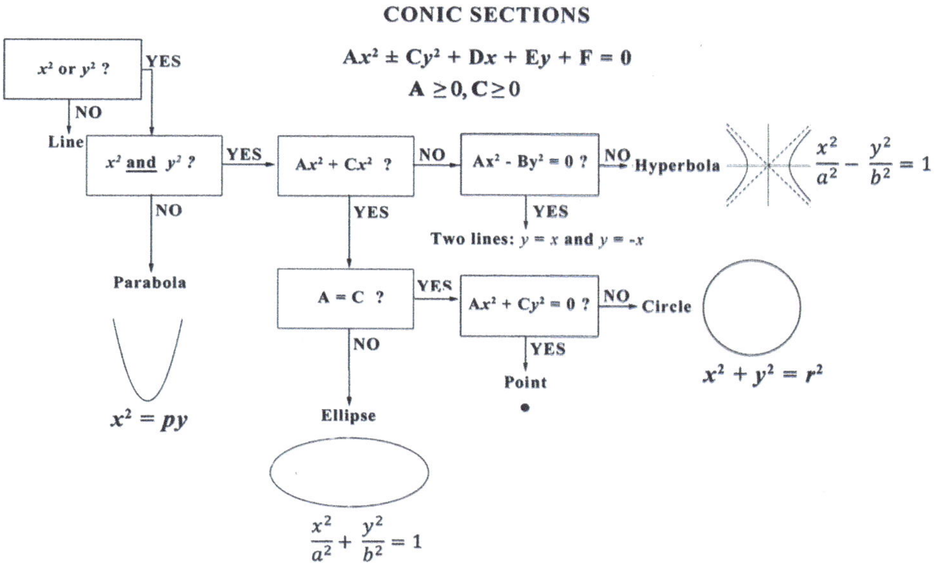

Anatomy of a hyperbola

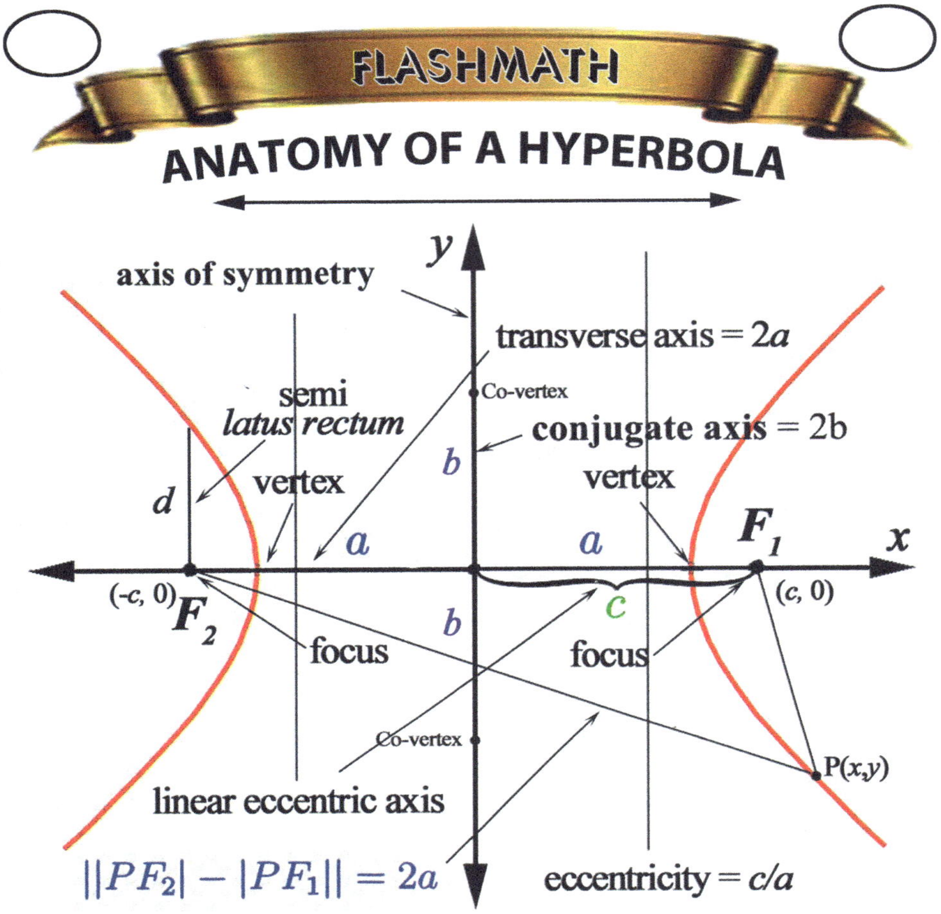

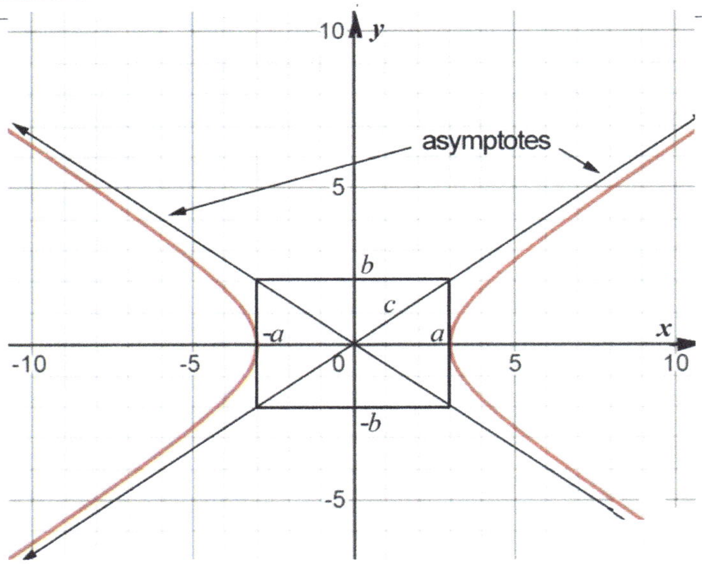

Anatomy of ellipse & parabola

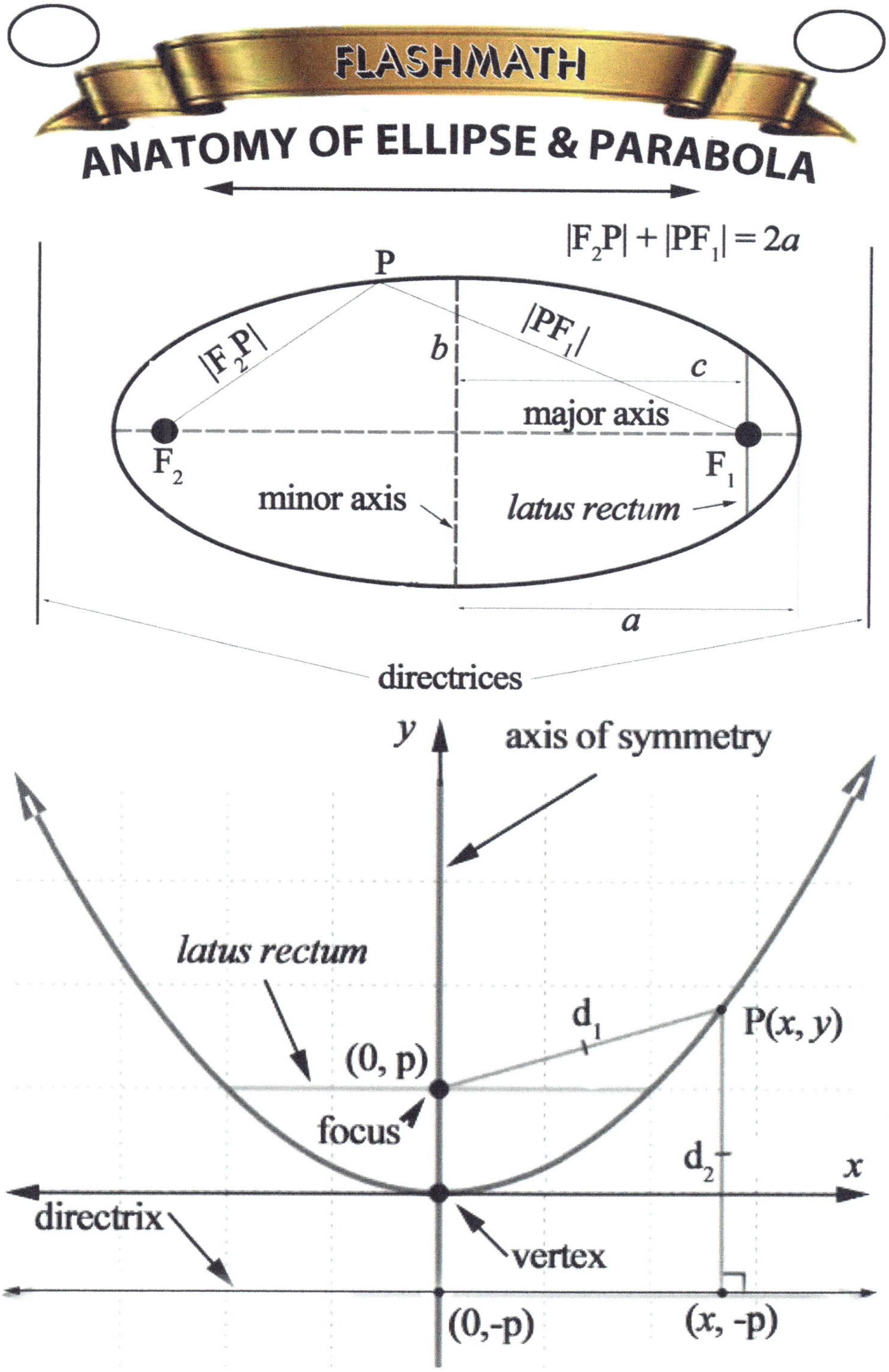

Analysis of a function worksheet

FLASHMATH

ANALYSIS OF A FUNCTION

ANALYSIS OF A FUNCTION WORKSHEET

Sketch Graph Here

f(x) =

Graphical Calculator
Window Size
[ ] by [ ]

Parent Function (if any):

$y = x$ $y = x^2$

$y = x^3$ $y = \sqrt{x}$

$y = |x|$ $y = 1/x$

$y = a^x$ $y = \log_a(x)$

$y = \sin(x)$ $y = \cos(x)$ $y = \tan(x)$

Comments:

Domain:

Range:

Roots: Real

Complex

Continuous on:

Discontinuities:

Jump

Removable ("hole")

Asymptotes: Horizontal

Vertical

Slant

Increasing on

Decreasing on

Symmetry (axis):

Bounded Below:

Bounded Above

Local Extrema: Maxima:

Minima:

End Behavior

As x approaches -∞,
f(x) approaches ______

As x approaches +∞,
f(x) approaches ______

Factoring the quadratic

FLASHMATH

FACTORING THE QUADRATIC

Factoring the Quadratic Equation $ax^2 + bx + c = 0$

By Inspection Employing the mn and m+n Technique

Simply answer the question "For the monic ($a = 1$) quadratic, $x^2 + bx + c$, what two integers, m and n, will result in $c =$ mn and $b =$ m + n?" When factoring, it is a good policy to spend a short time trying this method before proceeding to more complicated factoring techniques.

Complete the Square

$$ax^2 + bx + c = a(x - \mathrm{h})^2 + \mathrm{k},$$

where $\mathrm{h} = -b/2a$ and $\mathrm{k} = c - b^2/4a$.

Box (or Table) Method

The first step is to find two integers m and n such that mn $= ac$ and m + n $= b$. Next, construct a "box" or table to lay out the problem.

Factor by Grouping

This technique requires you to group terms with common factors before factoring. The usual approach is to "split" the middle term into two terms that will permit the factoring.

The Quadratic Formula

The formula *always works* for the quadratic of the form $ax^2 + bx + c = 0$. The two solutions are given by the quadratic formula:

$$x=(-b \pm\sqrt{(b^2 - 4ac)})/2a$$

The expression $b^2 - 4ac$ is the discriminant and provides much information about the underlying parabola. There are three cases:

1) $b^2 = 4ac$; the discriminant is zero, and the parabola's vertex touches the x-axis at a single value of x.

2) $b^2 > 4ac$; the discriminant is positive and the parabola crosses the x-axis at two different locations. Both solutions are real.

3) $b^2 < 4ac$; the discriminant is negative. The solutions have the form $x = u + iw$ and its conjugate $x = u - iw$. The parabola is above the x-axis.

Fibonacci sequences & palindromes

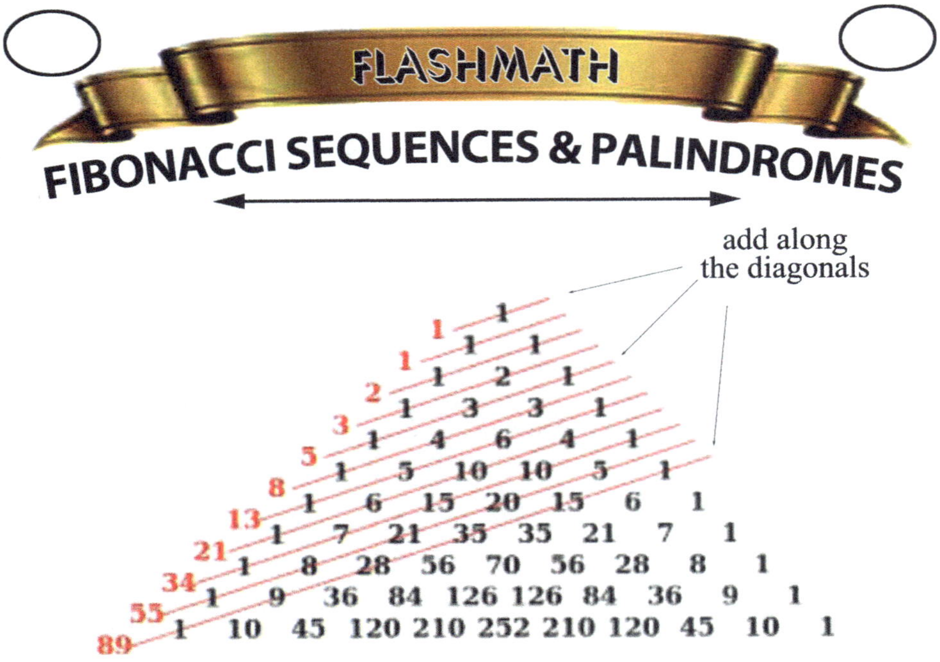

Fibonacci sequences in Pascal's Triangle

1^2	=	1
11^2	=	121
111^2	=	12321
11111^2	=	1234321
11111^2	=	123454321
111111^2	=	12345654321
1111111^2	=	1234567654321
11111111^2	=	123456787654321
111111111^2	=	12345678987654321
1111111111^2	=	1234567890098765 4321

MATH PALINDROMES

F.O.I.L. & other handy notes

FLASHMATH

F.O.I.L. & OTHER HANDY NOTES

FOIL

F O

$$(a + b)(c + d)$$

I L

$$= ac + ad + bc + bd$$

Quadratic Formula

The solution to the quadratic equation of the form

$ax^2 + bx + c = 0$

is

$$x = \frac{-b \pm \sqrt{b^2 - 4ac}}{2a}$$

Sum of Squares

$$a^2 + b^2 = (a + b)^2 - 2ab$$

Difference of Squares

$$a^2 - b^2 = (a + b)(a - b)$$

Perfect Square

$$(a + b)^2 = a^2 + 2ab + b^2$$

Sum of Cubes

$$a^3 + b^3 = (a + b)(a^2 - ab + b^2)$$

Difference of Cubes

$$a^3 - b^3 = (a - b)(a^2 + ab + b^2)$$

Linear equation forms

FLASHMATH

LINEAR EQUATION FORMS

SLOPE-INTERCEPT FORM	$y = mx + b$	(slope = m) (y-intercept = b)
POINT-SLOPE FORM	$y - y_1 = m(x - x_1)$	(slope = m)
TWO-POINT FORM	$y - y_1 = [(y_2 - y_1)/(x_2 - x_1)]\,(x - x_1)$	[slope = $(y_2 - y_1)/(x_2 - x_1)$]
SYMMETRIC FORM	$(x_2 - x_1)(y - y_1) = (y_2 - y_1)(x - x_1)$	[slope = $(y_2 - y_1)/(x_2 - x_1)$]
INTERCEPT FORM	$x/a + y/b = 1$ $a \neq 0$ and $b \neq 0$	(x-intercept = a; y-intercept = b)
GENERAL FORM	$Ax + By + C = 0$	(slope = -A/B; B ≠ 0)
STANDARD FORM	$Ax + By = C$	(slope = -A/B; B ≠ 0)

Note: For both the general and standard forms,
the x-intercept is -C/A if A ≠ 0, and the y-intercept is -C/B if B ≠ 0.
The slope of the line is -A/B if B ≠ 0.

Miscellaneous math notes

FLASHMATH

MISCELLANEOUS MATH NOTES

WHAT IS A SIGNED NUMBER?

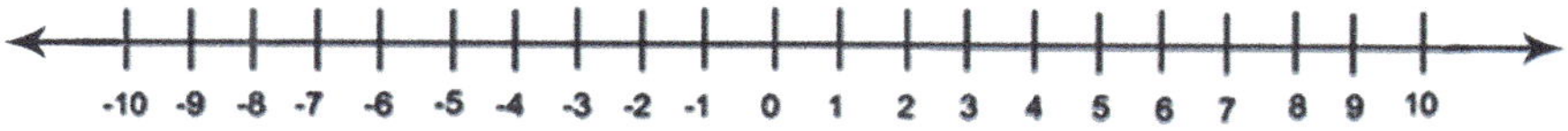

The figure above is called the *number line.* Notice that the number 0 is at the center. Now, negative numbers fall to the left of 0 (we say that a negative number is *less than zero*) and positive numbers fall to the right of 0 (we say that a positive number is *greater than zero*). That is, the sign of a number indicates whether it is to the left or right of zero. Notice also that any number, positive or negative, is simply its distance from zero.

IMPORTANT PATTERNS TO RECOGNIZE

Sum of squares	$a^2 + b^2 = (a+b)^2 - 2ab$
Difference of squares	$(a+b)(a-b) = a^2 - b^2$
Perfect square	$(a+b)^2 = a^2 + 2ab + b^2$
Sum of cubes	$a^3 + b^3 = (a+b)(a^2 - ab + b^2)$
Difference of cubes	$a^3 - b^3 = (a-b)(a^2 + ab + b^2)$
Pythagorean identity	$\sin^2(x) + \cos^2(x) = 1$
Euler's formula	$e^{i\theta} = \cos(\theta) + i\sin(\theta) = \text{cis}(\theta)$
Conjugate pairs	$a + i\,b \qquad a - i\,b$
	$a + \sqrt{b} \qquad a - \sqrt{b}$

USEFUL FORMULAS

Distance D between two points (x_1, y_1) and (x_2, y_2):

$$\sqrt{((x_1 - x_2)^2 + (y_1 - y_2)^2)}$$

Parent functions

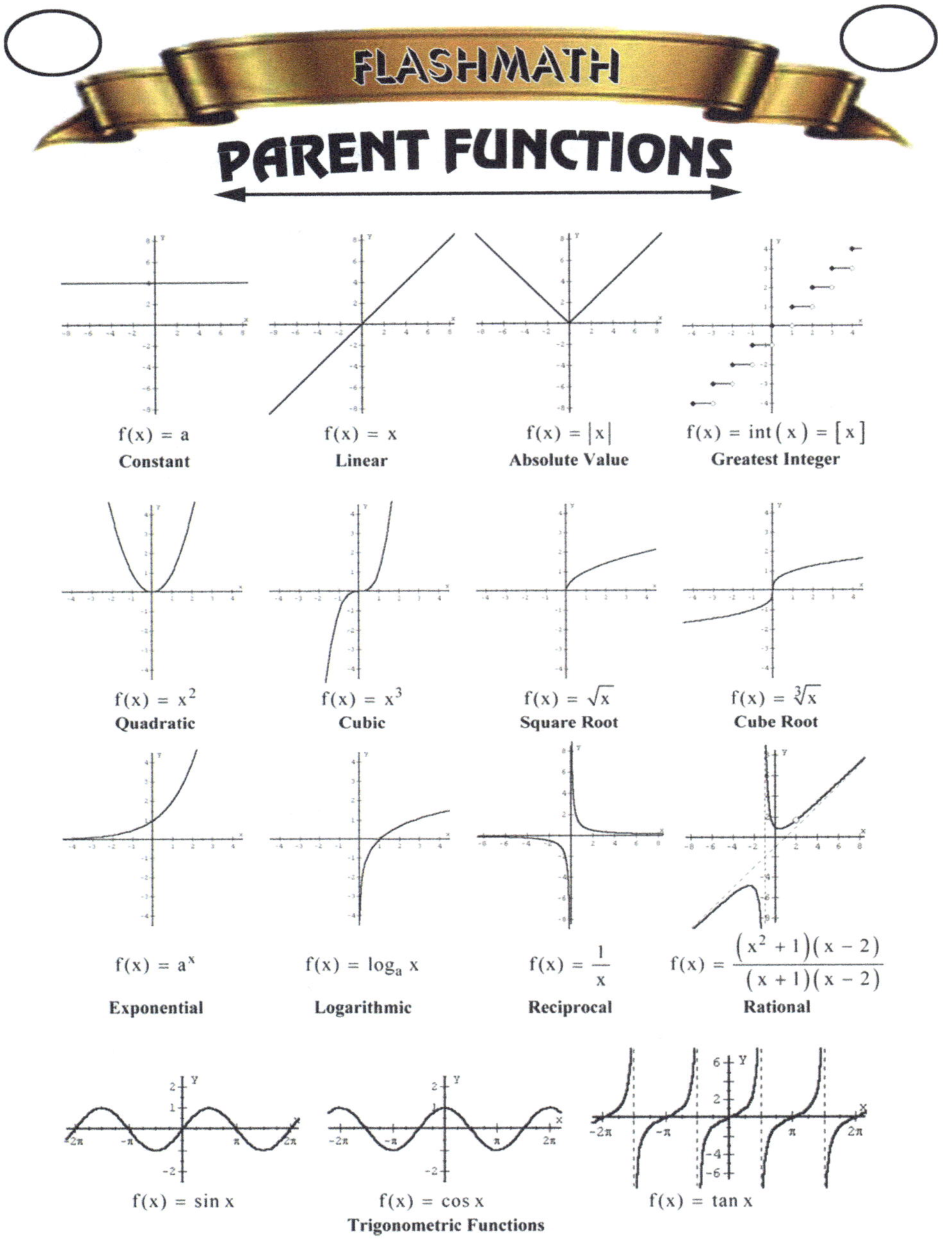

<u>Note</u>: The function $f(x - h) + k$ translates (moves) the parent function $f(x)$ "h" units to the right on the x-axis and "k" units up on the y-axis. If $h < 0$ and $k < 0$, the translation is left and down, respectively.

Pascal's triangle

FLASHMATH

PASCAL'S TRIANGLE

Pascal's triangle is a triangular array of numbers in which the top row consists of a single number, 1, and each subsequent row begins and ends with "1," and in any row, a number is the sum of the two adjacent numbers above it (blank equals zero). It is useful for binomial expansion, as shown below:

1	$(a+b)^0$
1 1	$(a+b)^1$
1 2 1	$(a+b)^2$
1 3 3 1	$(a+b)^3$
1 4 6 4 1	$(a+b)^4$
1 5 10 10 5 1	$(a+b)^5$
1 6 15 20 15 6 1	$(a+b)^6$
1 7 21 35 35 21 7 1	$(a+b)^7$

For example, expand the binomial $(a + b)^6$. Using the row 6 (the first row is row 0), 1 6 15 20 15 6 1, as the coefficients of the expanded term, we begin with a^6b^0, increasing the power of b and decreasing the power of a in each term, so that the sum of the powers of a an b is 6 for each term:

$$a^6b^0 + 6a^6b^1 + 15a^4b^2 + 20a^3b^3 + 15a^2b^4 + 6a^1b^5 + a^0b^6$$

which simplifies to

$$a^6 + 6a^6b + 15a^4b^2 + 20a^3b^3 + 15a^2b^4 + 6ab^5 + b^6.$$

Note that in the expansion of the binomial $(a + b)^n$ any entry in Pascal's Triangle is simply $\binom{n}{column\ number}$ which is the coefficient of $a^{(\text{n - column number})}\,b^{(\text{column number})}$. For example, in the expansion of $(a + b)^6$, the third term of the expansion from Row 6 is in Column 2 (remember, column count begins with 0).

Row 6 of Pascal's Triangle: 1 6 15 20 15 6 1

Third term in the in the expansion of $(a + b)^6$: $\binom{6}{2}a^{6-2}b^2 = 15a^4b^2$.

FLASHMATH

FEATURES OF PASCAL'S TRIANGLE

Pascal's Triangle left-justified

Col # →		0	1	2	3	4	5	6	7	8	9	10	11	12	
Row #			(1)	(2)	(3)										
0		1													
1		1	1												
2		1	2	1											
3	(4)	1	3	3	1										← $11^3 = 1331$
4		1	4	6	4	1									
5		1	5	10	10	5	1								
6	(5)	1	6	15	20	15	6	1							← Row sum = $2^6 = 64$
7		1	7	21	35	35	21	7	1						
8	(6)	1	8	28	56	70	56	28	8	1					← $(a + b)^8$ expansion coefficients
9		1	9	36	84	126	126	84	36	9	1				
10	(7)	1	10	45	120	210	252	210	120	45	10 (8)	1			
11		1	11	55	165	330	462	462	330	165	55	11	1		
12		1	12	66	220	495	792	924	792	495	220	66	12	1	

(1) **Natural numbers.** The "counting" numbers.

(2) **Triangular numbers.** The number of dots it takes to make various sized triangles.

(3) **Tetrahedral numbers, or triangular pyramidal numbers.** The number of dots needed to make pyramids with triangle bases.

(4) **Powers of 11.** The digits form the number of 11 raised to the power of the row number.

(5) **Powers of 2.** The sum of the row entries is 2 raised to the power of the row number.

(6) **Binomial expansion.** Each column entry is the coefficient of the term of the binomial expansion to the power of the row number.

(7) **Perfect squares.** The square of the number in column 1 is the sum of the two adjoining numberrs in column 2.

(8) **Combinations** $\binom{Row\ \#}{Column\ \#} = \binom{m}{n}$ where $0 \le n \le m$

(9) **Fibonacci numbers.** The sum of the numbers in the diagonal is the Fibonacci number in the sequence position of the column number.

PASCAL'S TRIANGLE & Fibonacci Numbers

Col # →	0	1	2	3	4	5	6	7	8	9	10	11	12	13
Row #		***1***	***1***	***2***	***3***	***5***	***8***	***13***	***21***	***34***	***55***	***89***	***144***	***233***
0	1													
1	1	1												
2	1	2	1											
3	1	3	3	1										
4	1	4	6	4	1									
5	1	5	10	10	5	1								
6	1	6	15	20	15	6	1							
7	1	7	21	35	35	21	7	1						
8	1	8	28	56	70	56	28	8	1					
9	1	9	36	84	126	126	84	36	9	1				
10	1	10	45	120	210	252	210	120	45	10	1			
11	1	11	55	165	330	462	462	330	165	55	11	1		
12	1	12	66	220	495	792	924	792	495	220	66	12	1	

Fibonacci sequence

(9)

(9) **Fibonacci numbers.** The sum of the numbers in the diagonal is the Fibonacci number in the sequence position of the column number.

Asymptotes

FLASHMATH

ASYMPTOTES

ASYMPTOTES

Let $f(x)/g(x)$ be a rational function where $f(x)$ is of degree n and $g(x)$ is of degree m.

•• Vertical asymptotes - If $g(a) = 0$ and $f(a) <> 0$, then a verti cal asymptote exists at $x = a$, and x is not in the domain of the rational function.

•• Horizontal asymptotes - there are three possible cases:

- **If $n < m$, then $y = 0$ is a horizontal asymptote;**
- **If $n > m$, then there is no horizontal asymptote; if $n = m + 1$, then there is a slant asymptote along the line determined by polynomial (or synthetic) division.**
- **If $n = m$, then $y = c$ is a horizontal asyptote where c is the quotient of the leading coefficients of $f(x)$ and $g(x)$, respectively.**

A function may cross a horizontal asymptote at most once.

•• A Hole (also discontinuity or removable singularity) occurs at that value of x where $f(x)$ is undefined. If a factor of $f(x)$ is identical to a factor of $g(x)$ then there is a hole at the value of x where the factor is equal to zero because there would result a factor of 0/0 of the rational function.

A *hole* differs from a *vertical asymptote* in that the latter occurs where $g(x) = 0$ and $f(x) <> 0$.

Change of base algorithm

FLASHMATH

CHANGE OF BASE ALGORITHM

CHANGE DECIMAL NUMBER TO BASE b < 10 (Example for b = 4)

Problem : Convert the decimal number N_{10} = 23,501 to N_4 base 4 (b = 4)

Line	*Base (b)*	N_{10}	*R*	*n*	$m=b^n$	*Check: R×m*
1	4	23,501				
2	4	5,875	1	0	1	1
3	4	1,468	3	1	4	12
4	4	367	0	2	16	0
5	4	91	3	3	64	192
6	4	22	3	4	256	768
7	4	5	2	5	1,024	2,048
8	4	1	1	6	4,096	4,096
9		0	1	7	16,384	16,384
				N_4 =	11233031	
					Total	23,501

Place number to be converted in Line 1, Column N_{10} . Set *i* = 1

1. **For Line *i* ...**
   Divide N_{10} by b, the new base (shown in Column "Base")
2. **Place the quotient in the next row (*I* + 1) of Column N_{10}**
3. **Place the remainder (including 0) in Column R of this row**
4. **Enter the number n = (Line number) - 2 (*e.g.*, for line 2, n = 0)**
5. **Calculate b^n and place in Column m**
6. **Calculate R × m and place product in Column "Check R × m"**
7. **If entry in Column N_{10} is 0, STOP; read the number in the new base from bottom to top of Column R (as shown above)**
   Otherwise, the value in Column N_{10} is not zero; increase *i* by 1 and repeat Steps 1 through 7 above.

Rules of exponents

FLASHMATH

RULES OF EXPONENTS

base → a^x ← exponent

Product rule:

(1) $x^a x^b x^c = x^{a+b+c}$

(2) $b^2a^3c^x = b^2a^3c^x$

(3) $b^2a^2c^2 = (bac)^2$

Quotient rule:

(1) $x^a/x^b = x^{a-b}$

(2) $b^2/a^3 = b^2/a^3$

(3) $b^2/(a^2c^2) = (b/ac)^2$

Powers of exponents rule

$(x^a)^b = x^{ab}$

Exponents and radicals rule

$x^{n/m} =$ *m*th root of x^n

Negative exponent rule

$x^{-a} = 1/x^a$; $1/x^{-b} = x^b$

Special cases of exponent rule

$n^0 = 1$ for $n \neq 0$

$0^n = 0$; $n>0$

$b^1 = b$

$1^n = 1$

$(-1)^n = 1$ if n is even or zero;

$(-1)^n = -1$ if n is odd.

Rules of logarithms

FLASHMATH
RULES OF LOGARITHMS

Product rule:

$$\log(xy) = \log(x) + \log(y)$$

Quotient rule:

$$\log(x/y) = \log(x) - \log(y)$$

Exponents rule:

$$\log(x^a) = a \log(x)$$

Change of base formula

$$\log_b(c) = \frac{\log_d(c)}{\log_d(b)}$$

New base

Old base

New base

Polynomial division

FLASHMATH

POLYNOMIAL LONG DIVISIION

Divide $3x^3 - 2x + 5$ by $(x - 1)$

Long Division

$$
\begin{array}{r l}
 & 3x^2 + 3x + 1 \\
\cline{2-2}
x - 1) & 3x^3 + 0x^2 - 2x + 5 \\
 & 3x^3 - 3x^2 \\
\cline{2-2}
 & \qquad 3x^2 - 2x + 5 \\
 & \qquad 3x^2 - 3x \\
\cline{2-2}
 & \qquad\qquad x + 5 \\
 & \qquad\qquad x - 1 \\
\cline{2-2}
 & \qquad\qquad\quad 6
\end{array}
$$

(MEMORIZE):

Let $\frac{f(x)}{g(x)}$ be a rational function where $f(x)$ is a polynomial of degree n and $g(x)$ is a polynomial of degree m.

"HOLES" If $f(x)$ and $g(x)$ have identical factors then a hole exists at each instance of the zero for the identical factors, *e.g.*, $(x - 2)/(x - 2)$ produces a hole at $x = 2$.

<u>Asymptote Rules</u>

– If $g(a) = 0$ and $f(a) \neq 0$, then $x = a$ is a <u>vertical</u> <u>asymptote</u>

– <u>Horizontal</u> <u>asymptotes</u> have 3 possible conditions:

- If $n < m$, then $y = 0$ is horizontal asymptote
- If $n > m$, then no horizontal asymptote
- If $n = m$, then $y = c$ is horizontal asymptote, where c is the quotient of leading coefficients of f and g

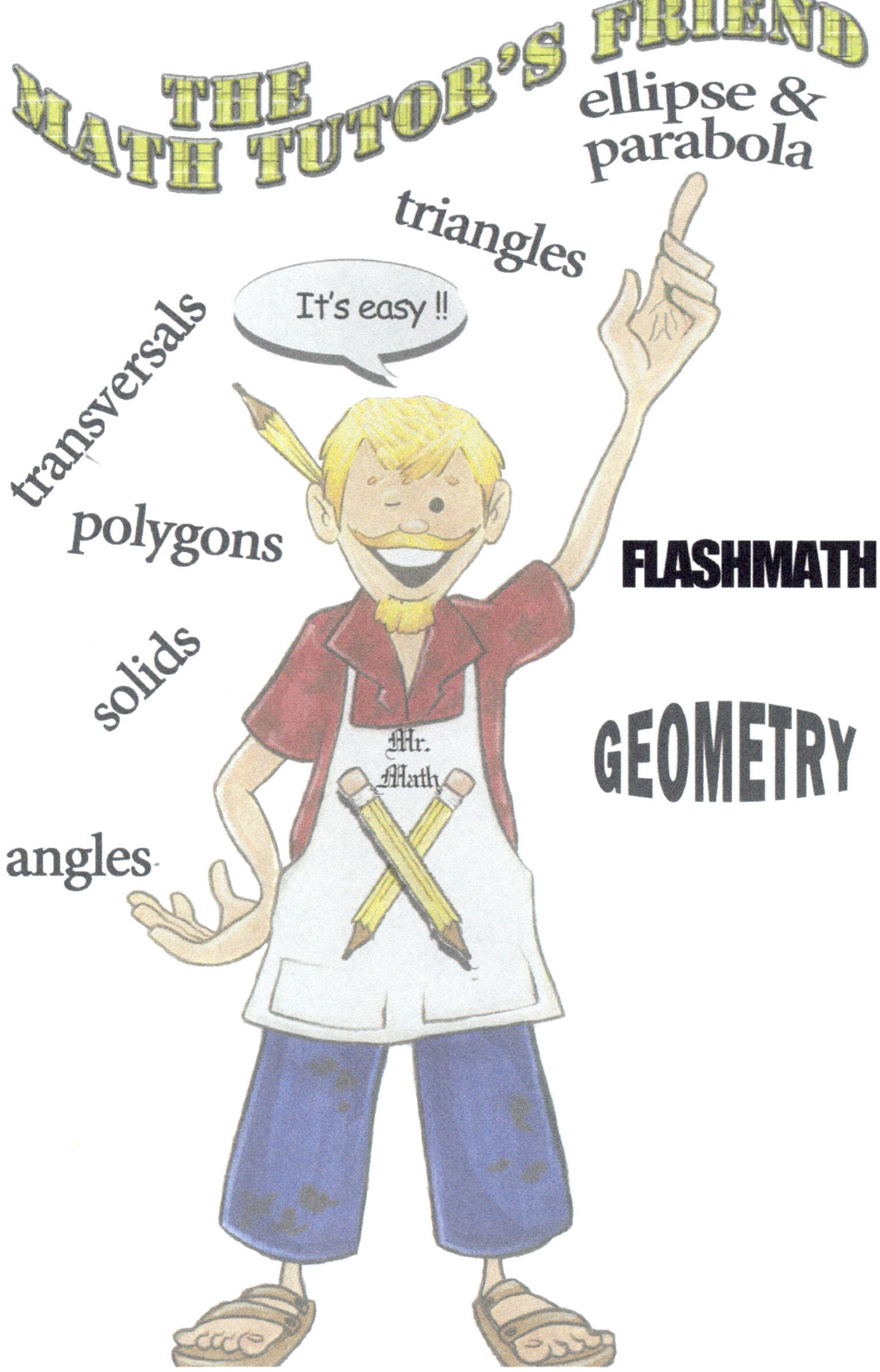
THE MATH TUTOR'S FRIEND
ellipse & parabola
triangles
It's easy !!
transversals
polygons
FLASHMATH
solids
GEOMETRY
Mr. Math
angles

Areas of polygons

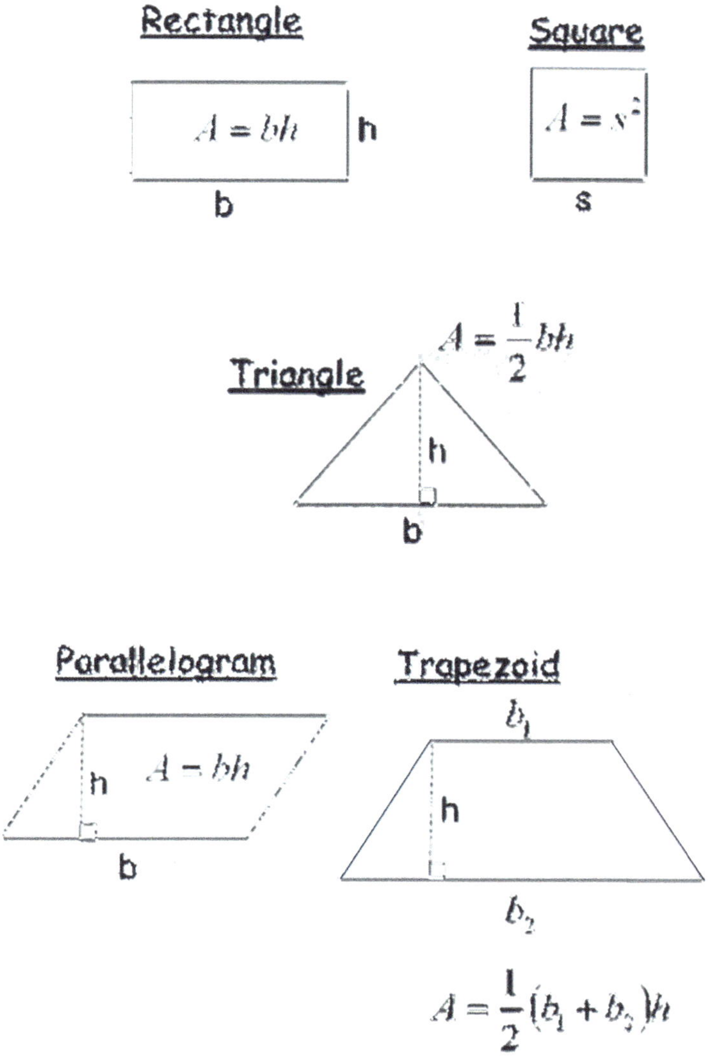

Complementary angles

FLASHMATH

COMPLEMENTARY ANGLES

(Need not be adjacent angles)

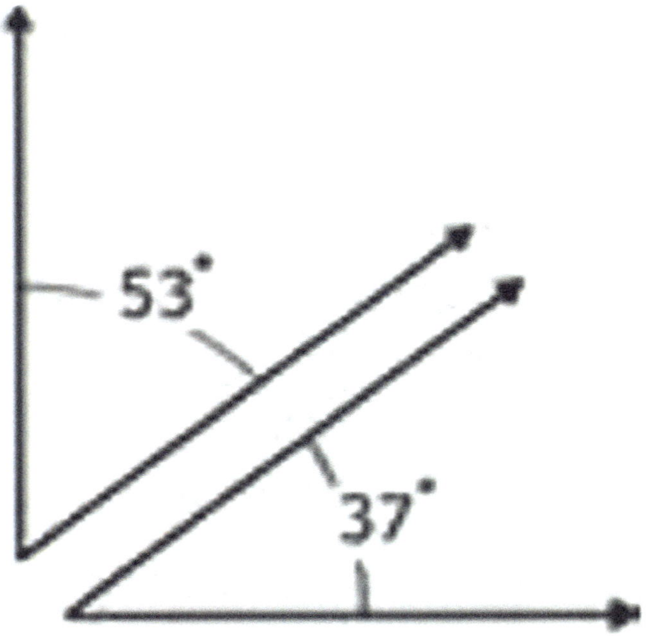

Cone

Surface area = $\pi \times r \times \sqrt{r^2 + h^2}$

Volume = $(1/3) \times \pi \times r^2 \times h$

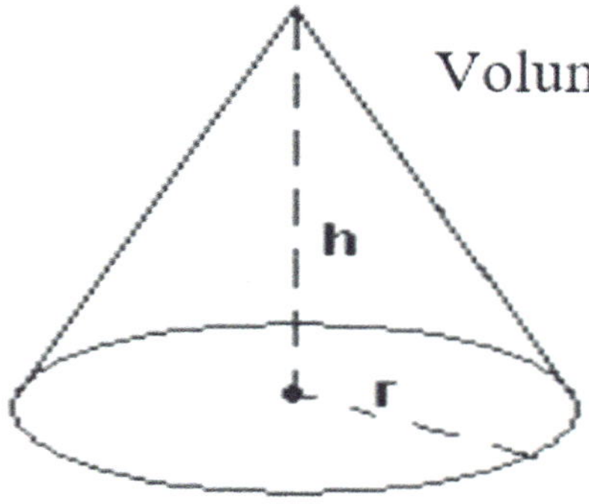

Cuboids

FLASHMATH

CUBOIDS

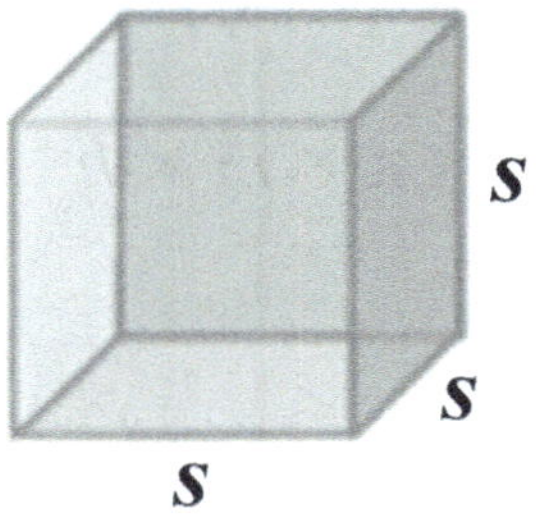

Surface Area $= 6s^2$
Volume $= s^3$

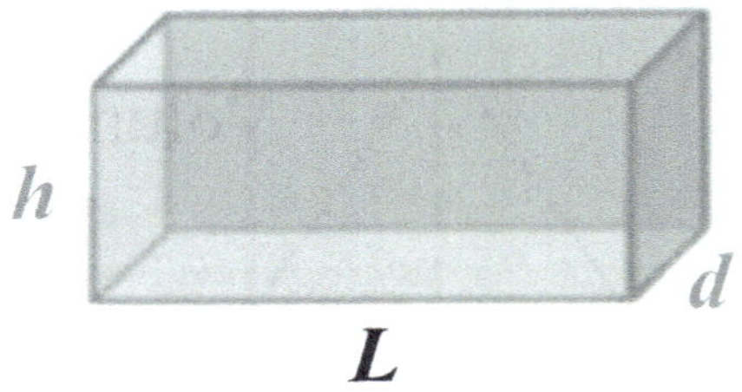

Surface Area $= 2 \times (Ld + hd + Lh)$
Volume $= h \times d \times L$

Right cylinder

Surface Area $= 2\pi r h + 2\pi r^2$

Volume $= V = \pi r^2 h$

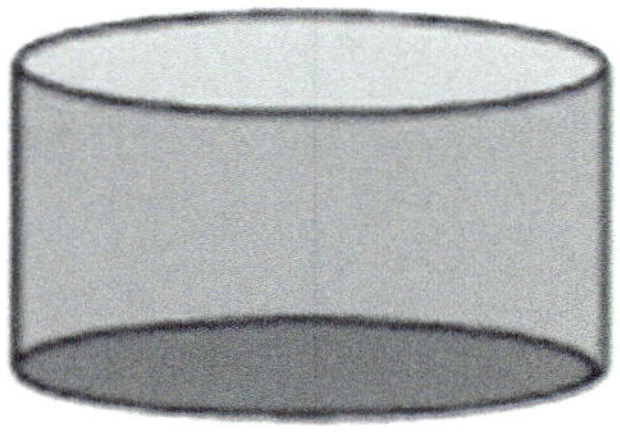

where

d = *diameter of cylinder*

r = *radius of cylinder*

h = *height of cylinder*

Right square pyramid

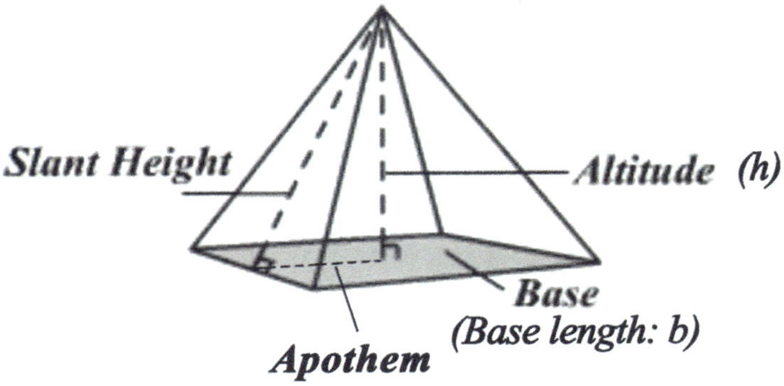

Base length b and height (altitude) h

Surface area $b^2 + b\sqrt{b^2 + 4h^2}$

Volume = $(1/3)\, b^2 h$

Solids

FEATURES OF SOME SOLIDS			
Solid	***Faces***	***Edges***	***Vertices (corners)***
cube	6	12	8
cuboid (box)	6	12	8
square pyramid	5	8	5
cylinder	3	2	0
cone	2	1	1
triangular prism	5	9	6
sphere	1	0	0
ellipsoid	1	0	0

Sphere

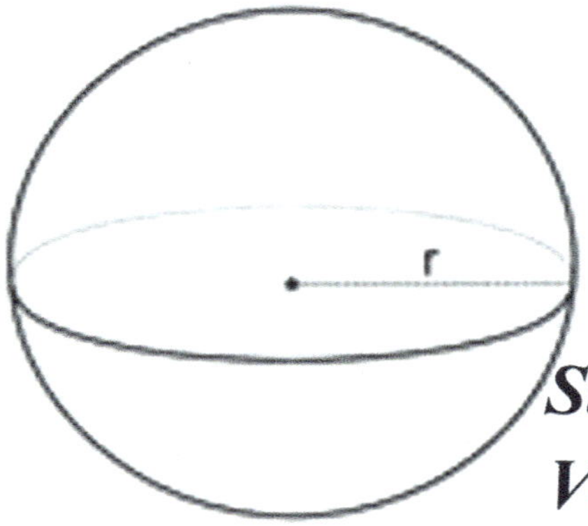

$$\textit{Surface area} = 4 \times \pi \times r^2$$

$$\textit{Volume} = (4/3) \times \pi \times r^3$$

Square & rhombus

FLASHMATH

SQUARE & RHOMBUS

Polygons
Quadrilaterals (4 sides)
Square and Rhombus

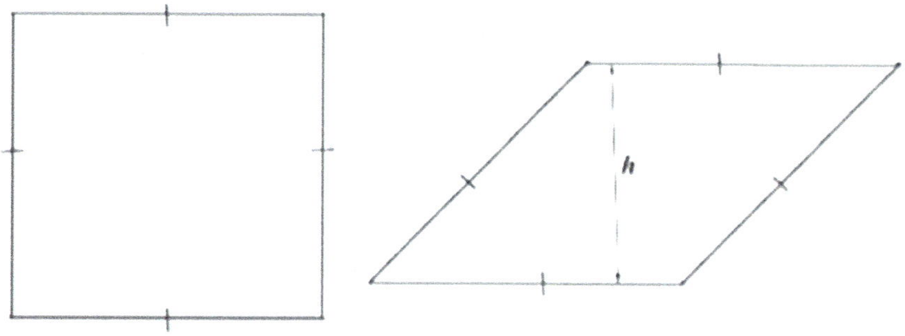

Matching "tick" marks signify congruence:
all sides have equal length

Triangles

FLASHMATH

TRIANGLES

The area of a triangle with vertices at coordinates (x_1, y_1), (x_2, y_2) and (x_3, y_3), is

$$\left|\frac{x_1(y_2 - y_3) + x_2(y_3 - y_1) + x_3(y_1 - y_2)}{2}\right|$$

Altitude from the right-angle vertex of a right triangle partions the triangle into two similar triangles

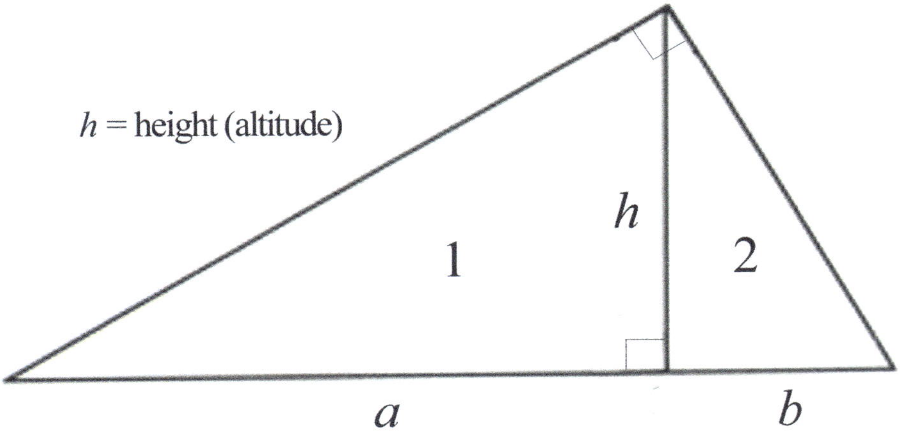

$$\triangle_1 \sim \triangle_2 \qquad \frac{b}{h} = \frac{h}{a} \qquad h = \sqrt{ab}$$

Triangular prism

FLASHMATH

TRIANGULAR PRISM

(isosceles ends)

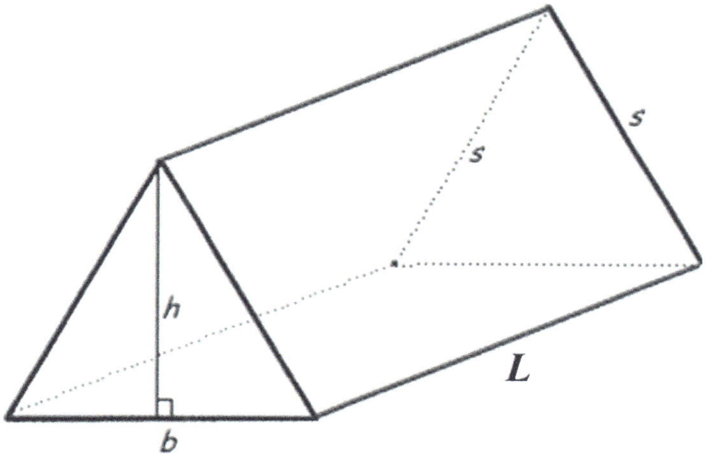

$$\textit{Surface Area} = b \times h + L \times (2s + b)$$

$$\textit{Volume} = \tfrac{1}{2} \times b \times h \times L$$

Types of angles

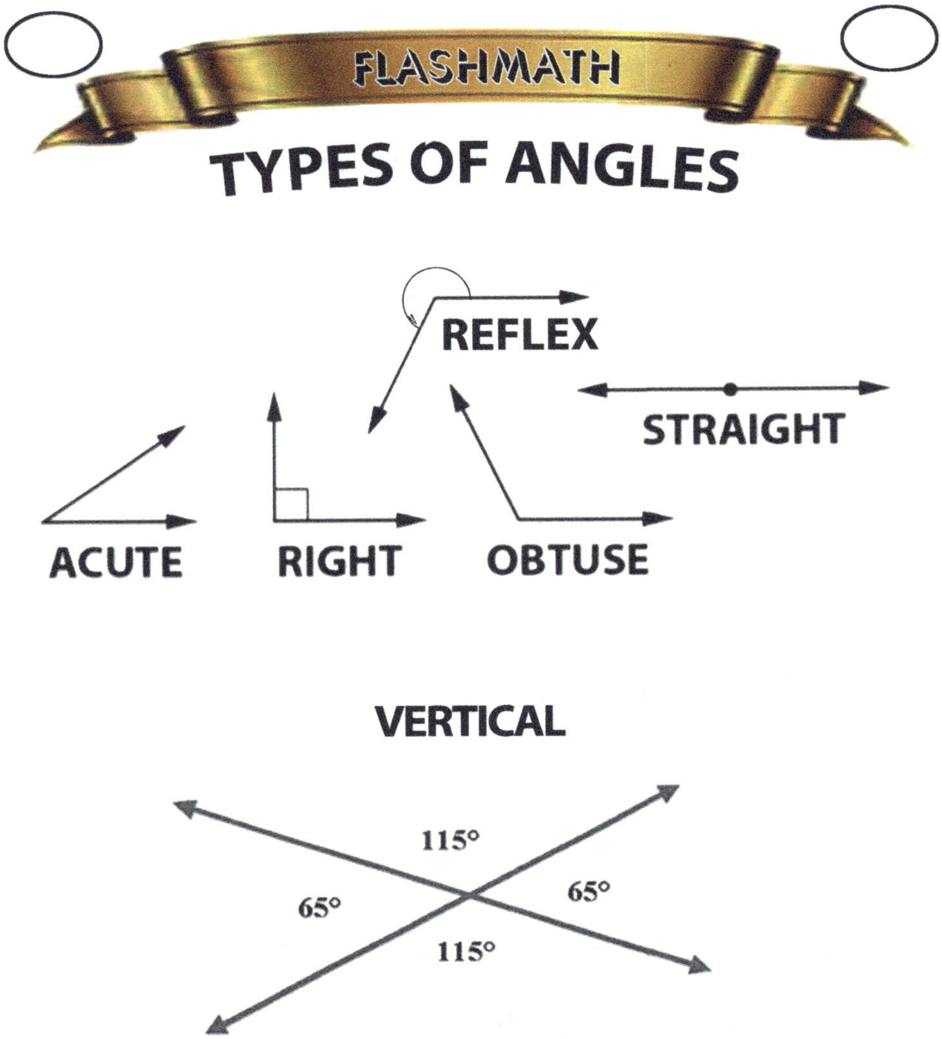

Note that adjacent angles

form "straight angles" (sum to 180°)

THE MATH TUTOR'S FRIEND
trig identities
dot product
Euler's formula
It's easy !!
sine & cosine
FLASHMATH
Mr. Math
unit circle
TRIGONOMETRY

Cross product properties

FLASHMATH

CROSS PRODUCT PROPERTIES

Properties of the cross product

For vectors u, v, and w and scalar *c*:

1. $\mathbf{u} \times \mathbf{v} = -\mathbf{v} \times \mathbf{u}$
2. $(c\mathbf{u}) \times \mathbf{v} = c(\mathbf{u} \times \mathbf{v}) = \mathbf{u} \times (c\mathbf{v})$
3. $\mathbf{u} \times (\mathbf{v} + \mathbf{w}) = \mathbf{u} \times \mathbf{v} + \mathbf{u} \times \mathbf{w}$
4. $(\mathbf{u} + \mathbf{v}) \times \mathbf{w} = \mathbf{u} \times \mathbf{w} + \mathbf{v} \times \mathbf{w}$
5. $\mathbf{u} \cdot (\mathbf{v} \times \mathbf{w}) = (\mathbf{u} \times \mathbf{v}) \cdot \mathbf{w}$
6. $\mathbf{u} \times (\mathbf{v} \times \mathbf{w}) = (\mathbf{u} \cdot \mathbf{w})\mathbf{v} - (\mathbf{u} \cdot \mathbf{v})\mathbf{w}$

The cross product is not associative

$$\mathbf{u} \times (\mathbf{v} \times \mathbf{w}) \neq (\mathbf{u} \times \mathbf{v}) \times \mathbf{w}$$

Dot product - area of a parallelogram

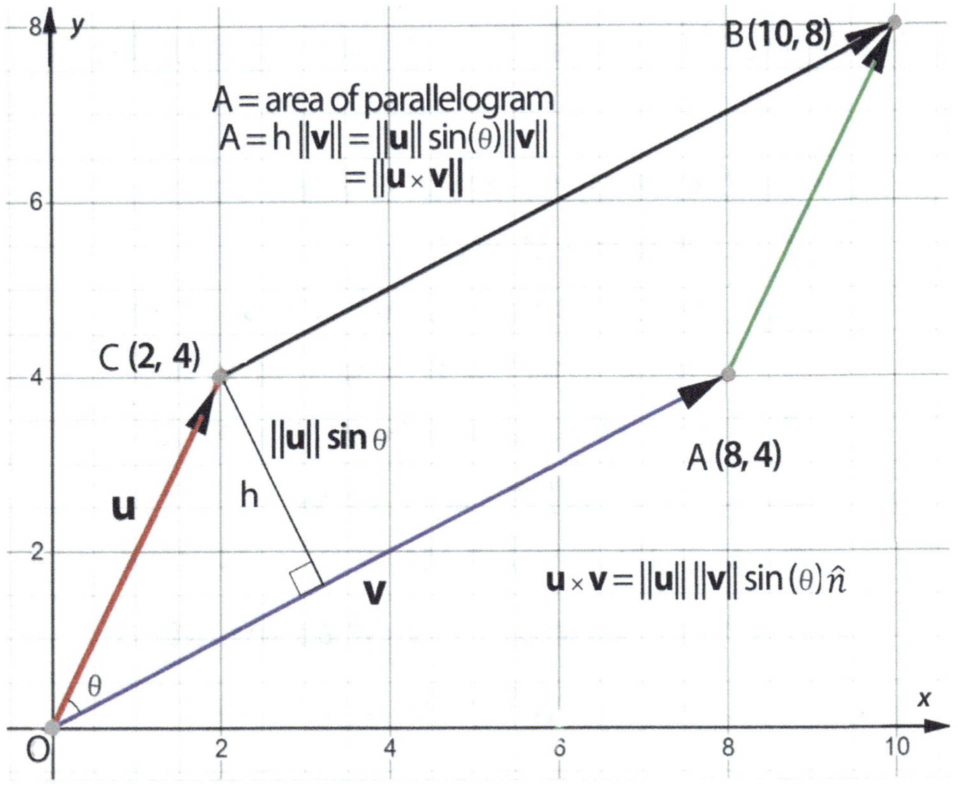

Dot product - U projected on V

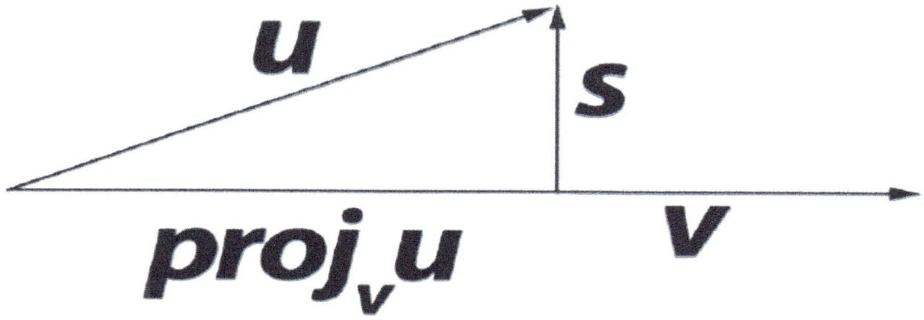

$$\mathbf{u} \cdot \mathbf{v} = \|\mathbf{u}\|\,\|\mathbf{v}\| \cos\theta = \frac{\|\mathbf{u}\|\,\|\mathbf{v}\|\,\|\text{proj}_{\mathbf{v}}\mathbf{u}\|}{\|\mathbf{u}\|}$$

$$= \|\mathbf{v}\|\,\|\text{proj}_{\mathbf{v}}\mathbf{u}\|$$

Dot product properties

FLASHMATH

DOT PRODUCT PROPERTIES

Properties of the dot product

1. $\mathbf{u}\cdot\mathbf{v} = \mathbf{v}\cdot\mathbf{u} = \|\mathbf{u}\|\,\|\mathbf{v}\|\cos\theta$ (commutative)
2. If $\mathbf{u}\cdot\mathbf{v} = 0$ then **u** or **v** (or both) are zero or **u** and **v** are orthogonal ($\cos 90° = 0$)
3. $(a\mathbf{u})\cdot(b\mathbf{v}) = (b\mathbf{u})\cdot(a\mathbf{v}) = ab\mathbf{u}\cdot\mathbf{v}$ (here, a and b are scalars)
4. $\mathbf{u}\cdot\mathbf{u} = \|\mathbf{u}\|^2$ (this means that $\boldsymbol{i}^2 = \boldsymbol{j}^2 = \boldsymbol{k}^2 = 1$)
5. $\mathbf{u}\cdot(\mathbf{v} + \mathbf{w}) = \mathbf{u}\cdot\mathbf{v} + \mathbf{u}\cdot\mathbf{w}$
6. If $\mathbf{u} = \langle u_1, u_2, u_3\rangle$ and $\mathbf{v} = \langle v_1, v_2, v_3\rangle$ then $\mathbf{u}\cdot\mathbf{v} = u_1v_1 + u_2v_2 + u_3v_3$

Euler's formula

FLASHMATH

EULER'S FORMULA

"A most beautiful equation"

An important trigonometric identity from the Swiss mathematician *Leonhard Euler* (1709-1783):

$$e^{i\theta} = cos(\theta) + isin(\theta) = cis(\theta)$$

Let $\theta = \pi$

$$e^{i\pi} = cos(\pi) + isin(\pi) = -1$$

Add 1 to both sides to get *Euler's equation* that relates $\boldsymbol{i}$, $\boldsymbol{e}$, $\boldsymbol{\pi}$, 1, and 0:

$$\mathbf{e^{i\pi} + 1 = 0}$$

where

- e is Euler's number, the base of natural logarithms,
- i is the imaginary unit, which satisfies $i^2 = -1$, and
- π is *pi*, the ratio of the circumference of a circle to its diameter.

Note the deep mathematical beauty! The basic arithmetic operations of addition, multiplication, and exponentiation occur exactly once.

The equation also links five mathematical constants:

$$\mathbf{0,\ 1,\ \pi,\ e;} \text{ and } \boldsymbol{i},$$

the imaginary unit of a complex number, defined as $i^2 = -1$.

•

Laws of sines - cosines - tangents

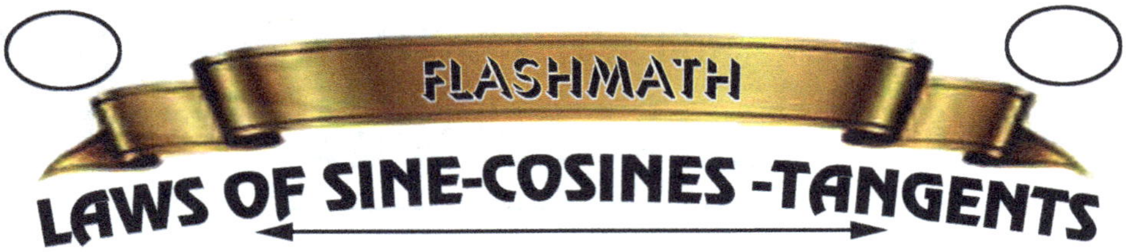

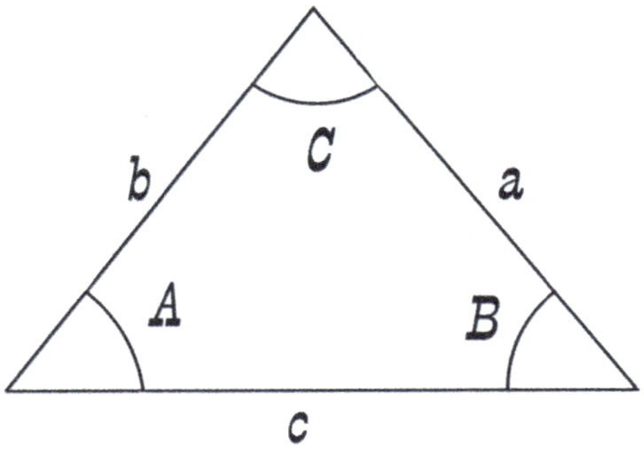

LAW OF SINES

The angles and their opposite sides of any triangle (see figure) are related by the Law of Sines:

$$\text{Sin}(A)/a = \sin(B)/b = \sin(C)/c$$

LAW OF COSINES

Referring to the triangle in the figure, the Law of Cosines states:

$$c^2 = a^2 + b^2 - 2ab\cos(C)$$

LAW OF TANGENTS

There is a relationship between the tangents of two angles of a triangle and the lengths of the opposing sides.
Let a, b, and c be the lengths of the three sides of the triangle, and A,B, and C be the angles opposite those three respective sides (see the figure). The law of tangents states that

$$(a - b)/(a + b) = \tan(1/2\ (A - B)) / \tan(1/2\ (A + B))$$

Sine function

FLASHMATH

SINE FUNCTION

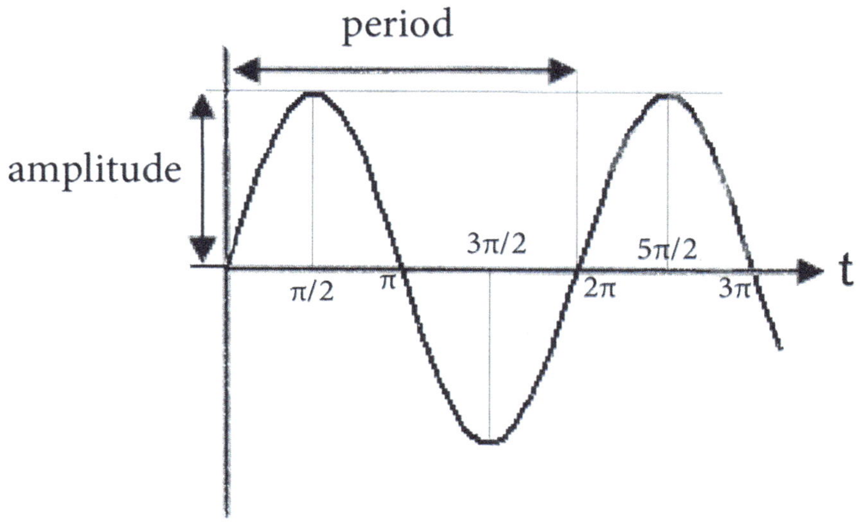

amplitude = |a|

$$f(t) = a \sin(\omega t)$$

$$\omega = 2\pi f$$

$$f = \text{freq} = \omega / 2\pi$$

Sine function nomenclature

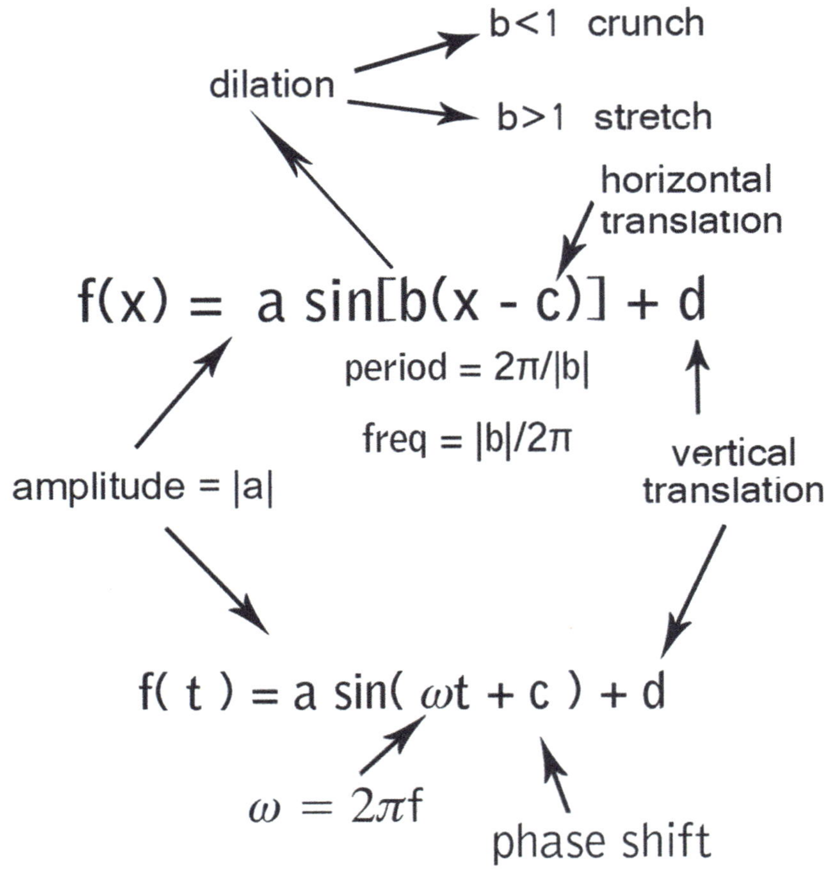

Trigonometric identities

FLASHMATH

TRIGONOMETRIC IDENTITIES

Pythagorean Identities

$$\sin2(\theta) + \cos2(\theta) = 1.$$
$$\tan2(\theta) + 1 = \sec2(\theta),$$
$$1 + \cot2(\theta) = \csc2(\theta).$$

Negative angle identities

$$\sin(-\theta) = -\sin(\theta)$$
$$\cos(-\theta) = \cos(\theta)$$

Sum of angles identities

$$\cos(\alpha+\theta) = \cos(\alpha)\cos(\theta) - \sin(\alpha)\sin(\theta)$$
$$\sin(\alpha+\theta) = \cos(\theta)\sin(\alpha) + \sin(\alpha)\cos(\theta)$$
$$\cos(\alpha-\theta) = \cos(\alpha)\cos(\theta) + \sin(\alpha)\sin(\theta)$$
$$\sin(\alpha-\theta) = \sin(\alpha)\cos(\theta) - \cos(\alpha)\sin(\theta).$$

De Moivre's formula

$$(\cos(\theta) + i\sin(\theta))n = \cos(n\theta) + i\sin(n\theta)$$

This identity holds only when "n" is an integer.

The formula can also be written as

$$(\text{cis}(\theta))n = \text{cis}(n\theta).$$

Unit circle

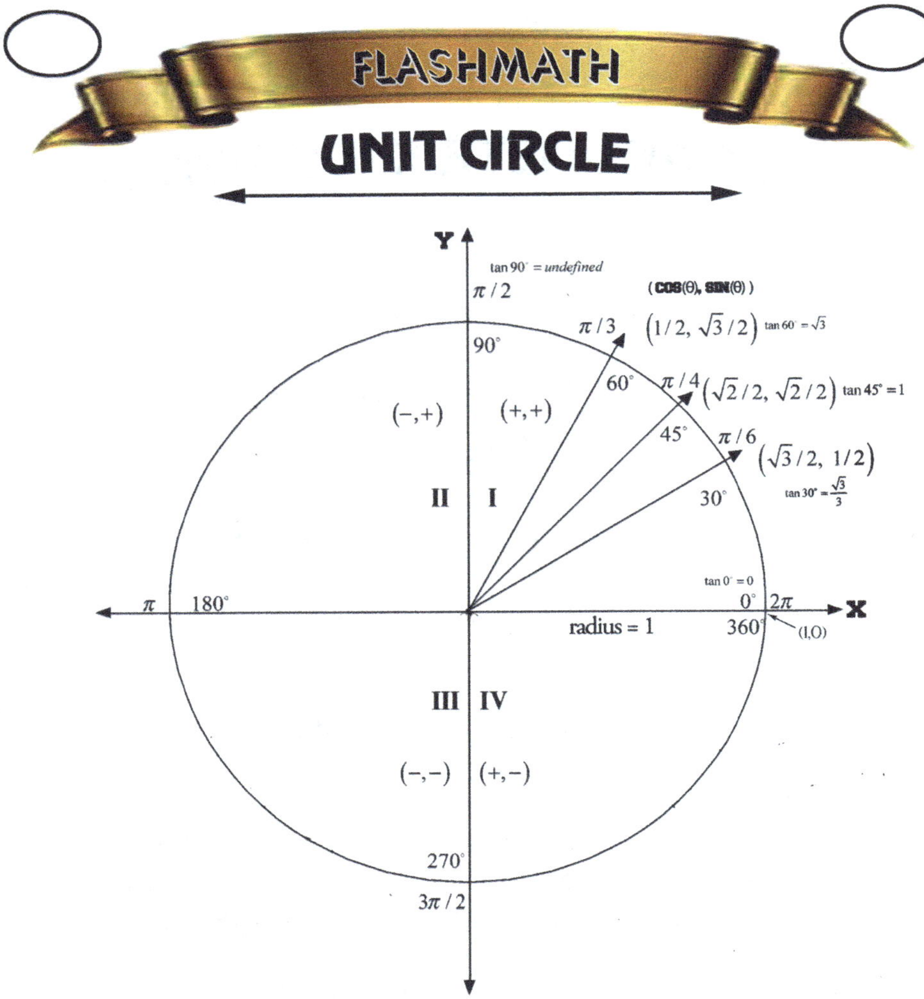

Anatomy of an Angle in Standard Position

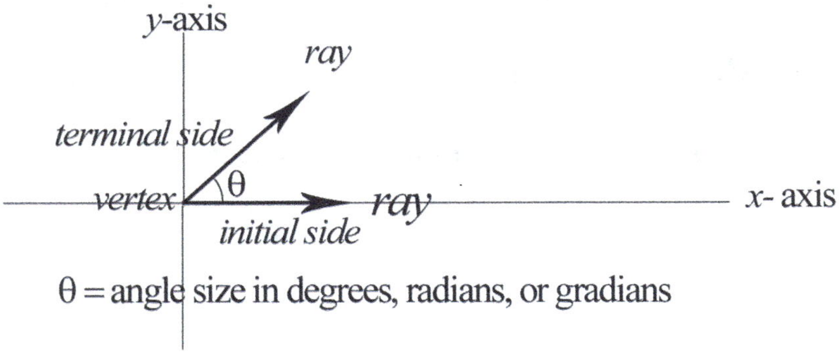

Vectors - triple products

FLASHMATH

VECTORS - TRIPLE PRODUCTS

Let u, v, and w, be vectors

scalar triple product

$$\mathbf{u}\cdot(\mathbf{v}\times\mathbf{w}) = \mathbf{v}\cdot(\mathbf{w}\times\mathbf{u}) = \mathbf{w}\cdot(\mathbf{u}\times\mathbf{v})$$

The absolute value of the scalar triple product is the volume of a parallelepiped defined by u, v, and w.

vector triple product

$$\mathbf{u}\times(\mathbf{v}\times\mathbf{w}) = (\mathbf{u}\cdot\mathbf{w})\,\mathbf{v} - (\mathbf{u}\cdot\mathbf{v})\,\mathbf{w}$$

Complex plane & complex conjugate

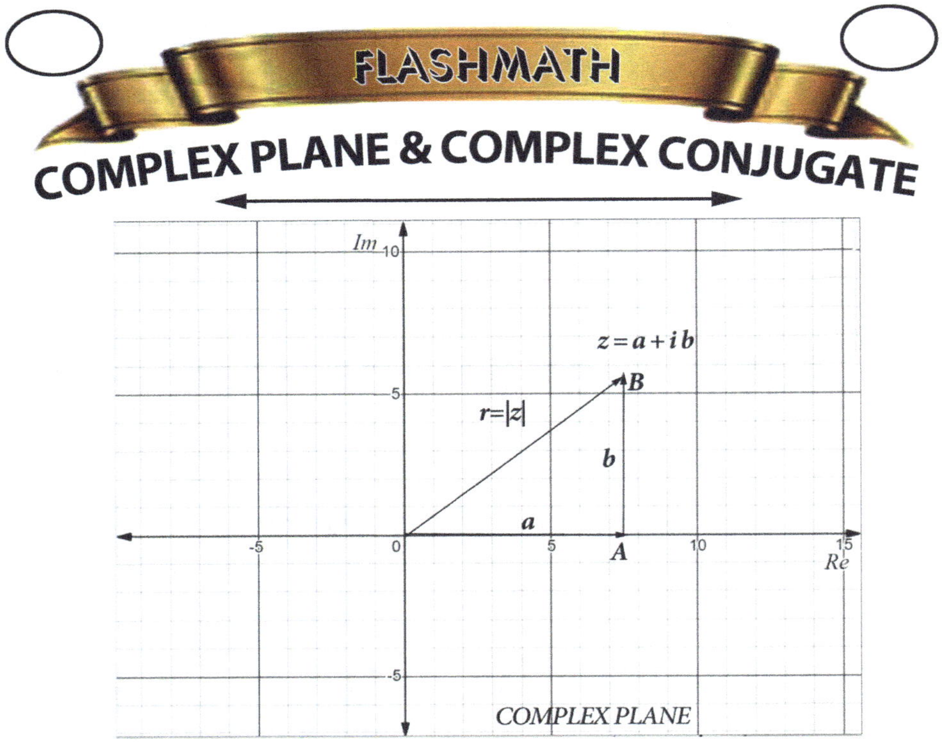

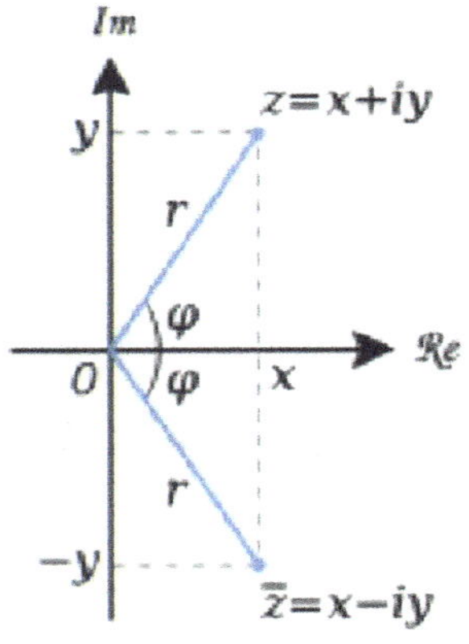

The *absolute value of a complex number z* is the distance *r* of *z* from the origin. As the figure shows, *z* and its *complex conjugate* have the same absolute value, *r*.

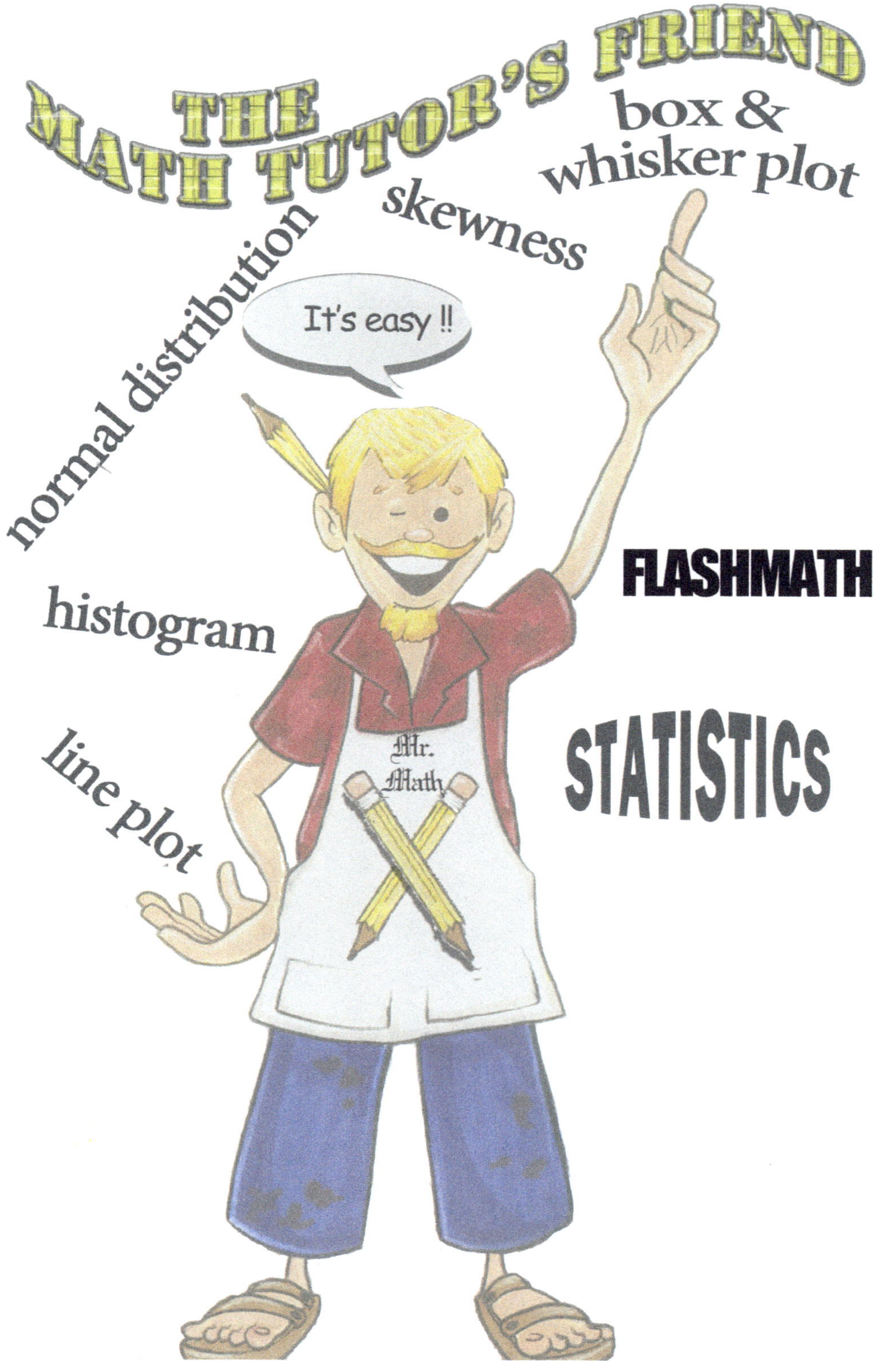
THE
MATH TUTOR'S FRIEND
box &
whisker plot
skewness
normal distribution
It's easy !!
FLASHMATH
histogram
Mr.
Math
STATISTICS
line plot

Box & whisker plot

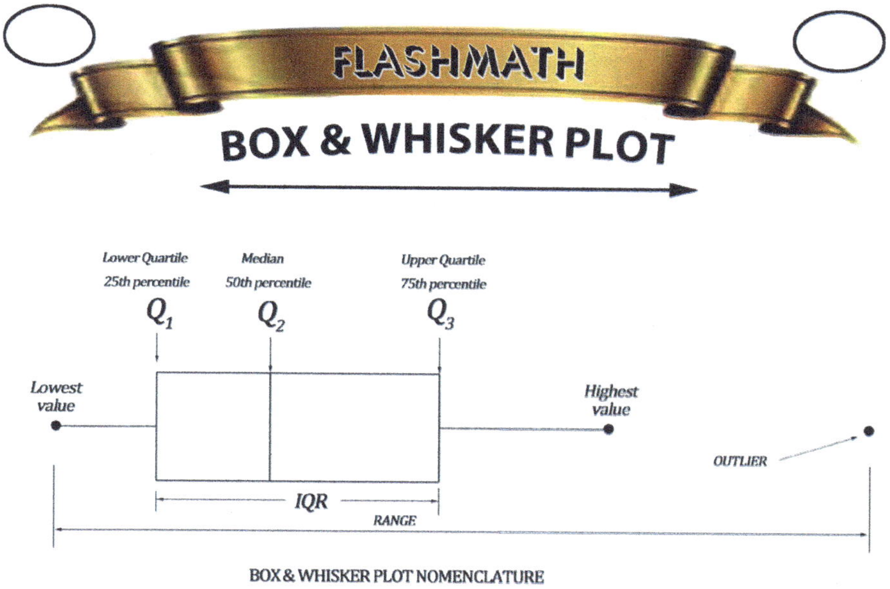

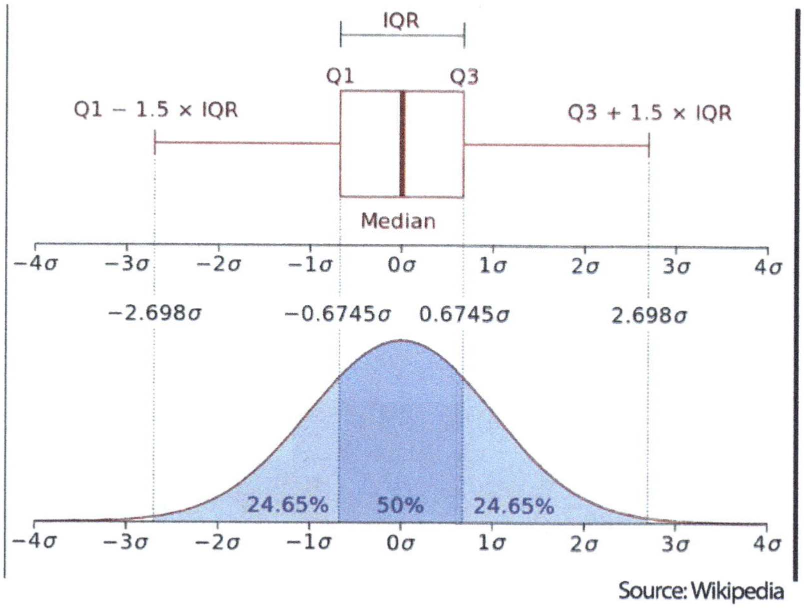

Line plot & histogram

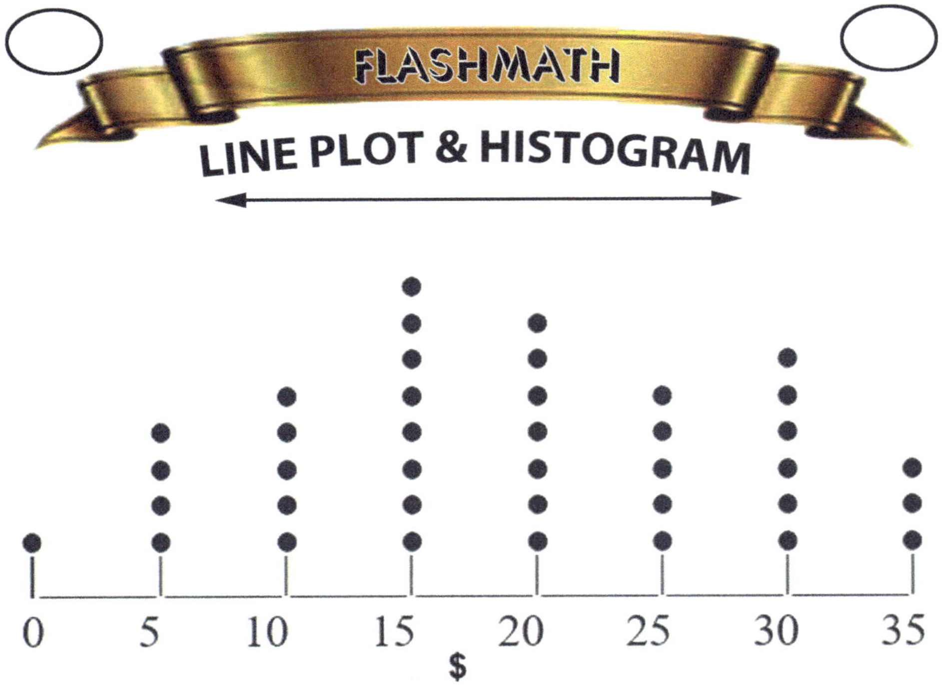

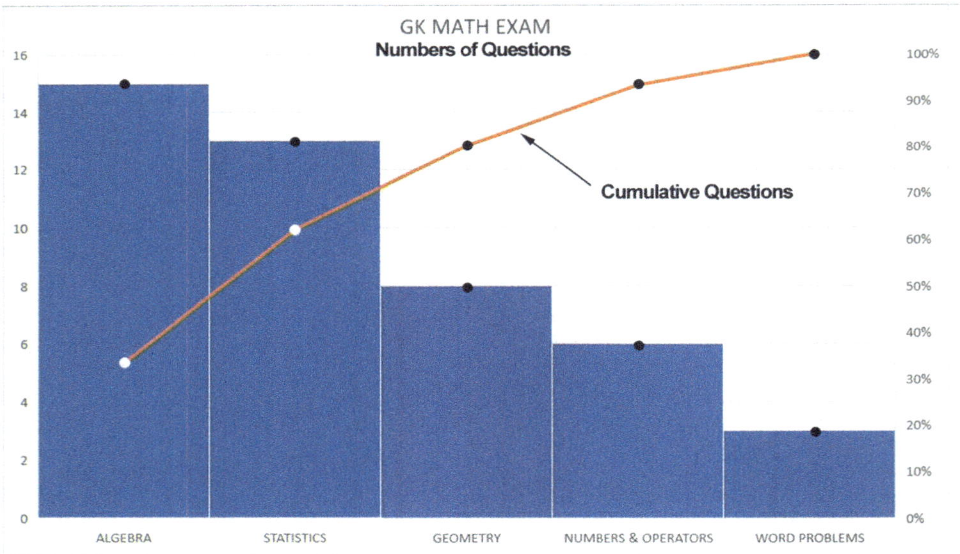

Logic & math symbols

FLASHMATH

LOGIC & MATH SYMBOLS

SYMB	NAME/MEANING	EXAMPLE
$\vee$	Disjunction	$p \vee q$ p or q (or both)
$\veebar$	Exclusive Disjunction	$p \veebar q$ p or q (but not both)
$\wedge$	Conjunction	$p \wedge q$ p and q
~	Negation	~p the statement p is False
$\rightarrow$	Conditional	$p \rightarrow q$ if p then q
$\Rightarrow$	Conditional (*e.g.*, of complex statements)	$p \wedge q \Rightarrow q$
$\leftrightarrow$	Biconditional; also abbreviated as "iff"	$p \leftrightarrow q$ p if and only if q
$\Leftrightarrow$ or $\equiv$	Equivalence (*e.g.*, of complex statements)	$p \wedge q \Leftrightarrow q \wedge p$
	QUANTIFIERS	
$\in$	"is an element (or member) of	"2 $\in$ {2,4,6,8,…}
$\ni$	"such that"	quantifier
$\exists$	"there exists"	quantifier
$\forall$	"for all"	$f(x) = g(x)\ \forall x$
$\therefore$	"Therefore"	
Q.E.D.	*Quod Erat Demonstrandum* "the proof is complete"	

Normal distribution & skewness

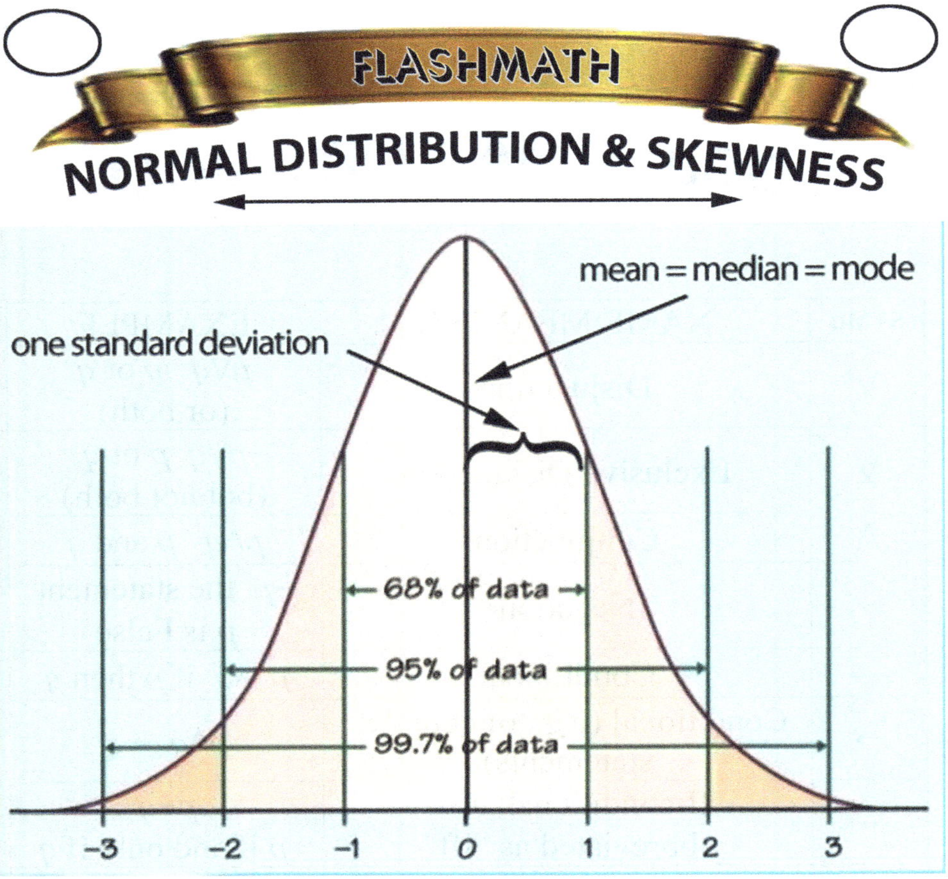

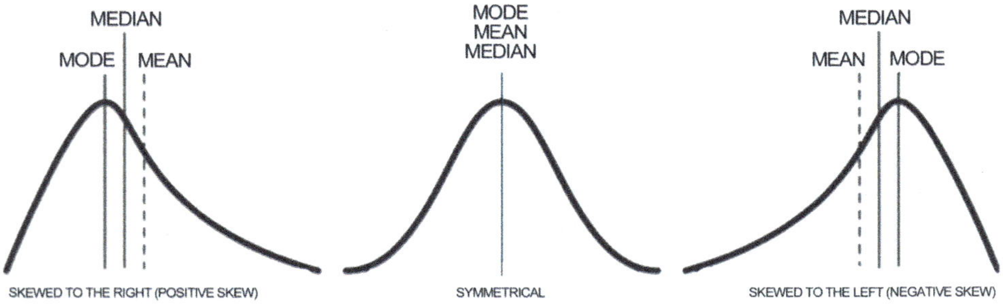

SKEWNESS OF PROBABILITY DISTRIBUTIONS

Rules of inference - 1

FLASHMATH

RULES OF INFERENCE - 1

Addition Law $p \Rightarrow p \vee q$

Associative Law $p \vee (q \vee r) \Leftrightarrow (p \vee q) \vee r$

$p \wedge (q \wedge r) \Leftrightarrow (p \wedge q) \wedge r$

Commutative Law $p \wedge q \Leftrightarrow q \wedge p$

$p \vee q \Leftrightarrow q \vee p$

Constructive dilemma

$(p \rightarrow q) \wedge (r \rightarrow s) \Rightarrow (p \vee r) \rightarrow (q \vee s)$

$(p \rightarrow q) \wedge (r \rightarrow s) \Rightarrow (p \wedge r) \rightarrow (q \wedge s)$

Contraposition $(p \rightarrow q) \Leftrightarrow (\sim p \rightarrow \sim q)$

De Morgan's Law $\sim (p \wedge q) \Leftrightarrow (\sim p \vee \sim q)$

$\sim (p \vee q) \Leftrightarrow (\sim p \wedge \sim q)$

Destructive dilemma

$(p \rightarrow q) \wedge (r \rightarrow s) \Rightarrow (\sim q \vee \sim s) \rightarrow (\sim p \vee \sim r)$

$(p \rightarrow q) \wedge (r \rightarrow s) \Rightarrow (\sim p \wedge \sim s) \rightarrow (\sim p \wedge \sim s)$

Disjunctive Syllogism $(p \vee q) \wedge \sim p \Rightarrow q$

Rules of inference - 2

FLASHMATH

RULES OF INFERENCE - 2

Distributive Law

$p \wedge (q \vee r) \Leftrightarrow (p \wedge q) \vee (p \wedge r)$

$p \vee (q \wedge r) \Leftrightarrow (p \vee q) \wedge (p \vee r)$

Double Negative $\sim(\sim p) \Leftrightarrow p$

Modus Ponens $(p \rightarrow q) \wedge p \Rightarrow q$

Modus Tollens $(p \rightarrow q) \wedge \sim q \Rightarrow \sim p$

Reductio ad Absurdum

$(p \rightarrow q) \Leftrightarrow (p \wedge \sim q) \rightarrow (q \wedge \sim q)$

Simplification $(p \wedge q) \Rightarrow p$

$(p \wedge q) \Rightarrow q$

Transitivity $(p \rightarrow q) \wedge (q \rightarrow r) \Leftrightarrow (p \rightarrow r)$

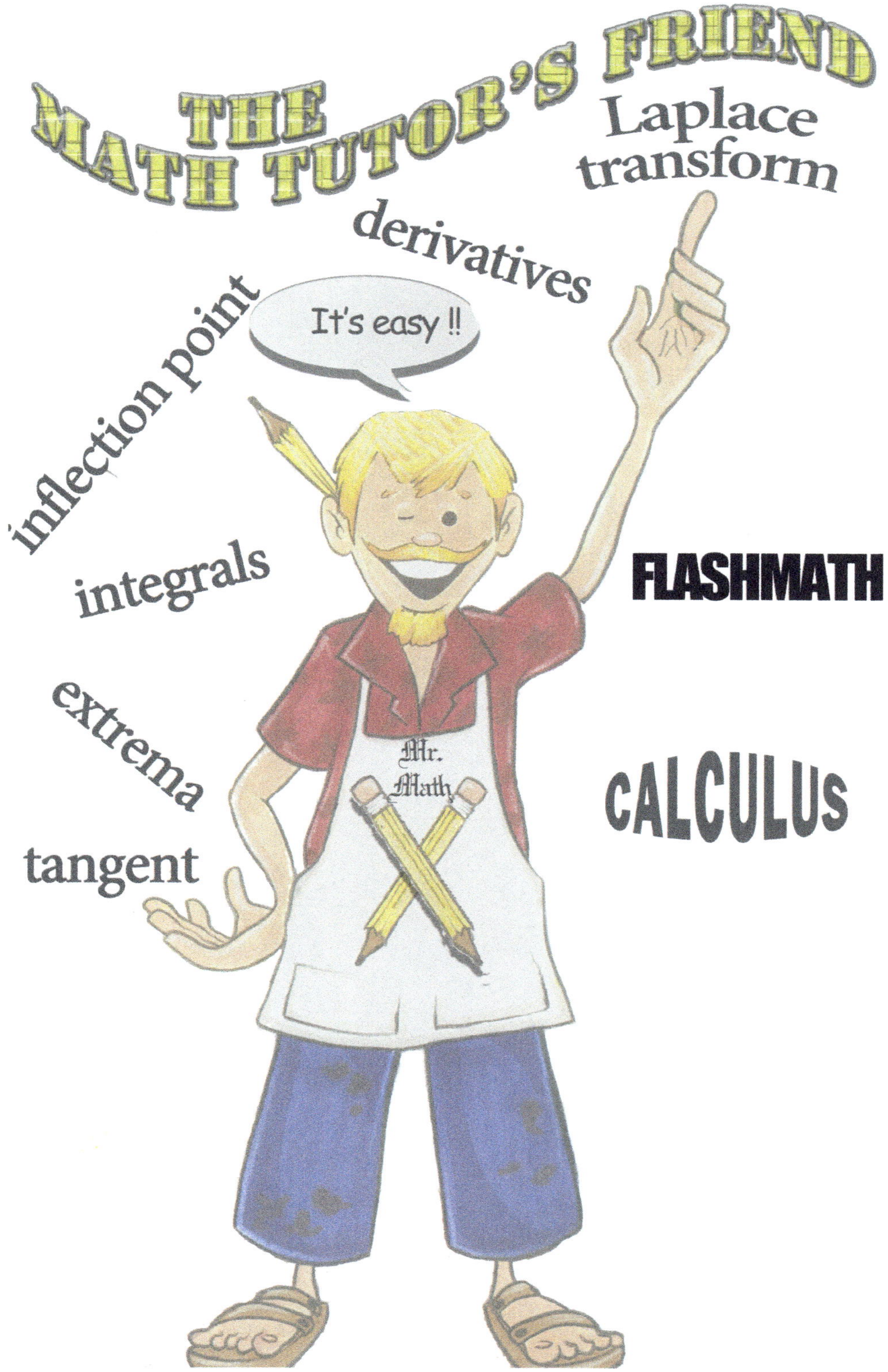
THE MATH TUTOR'S FRIEND
Laplace transform
derivatives
inflection point
It's easy !!
integrals
FLASHMATH
extrema
Mr. Math
CALCULUS
tangent

Anatomyof integrals

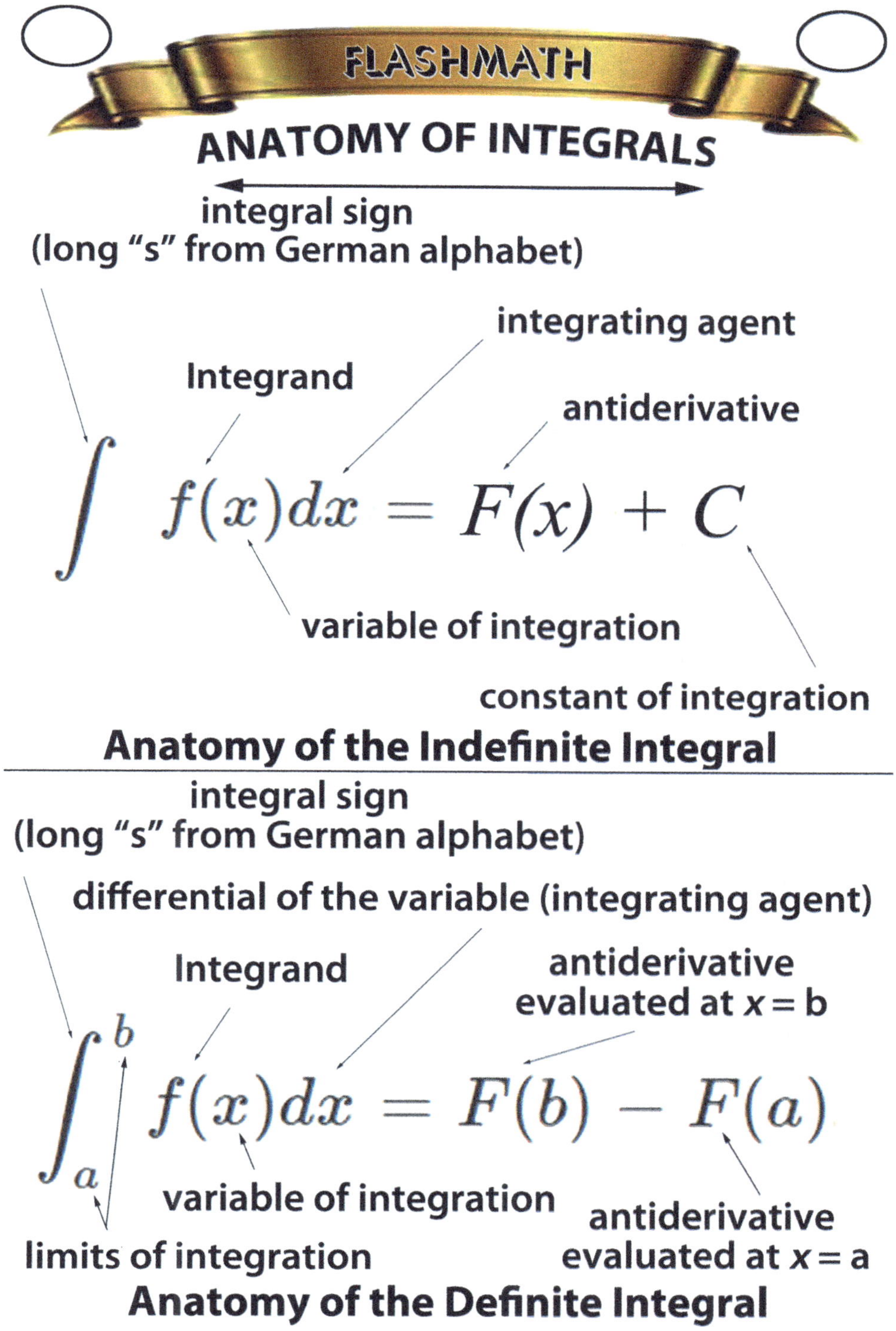

Definite integral

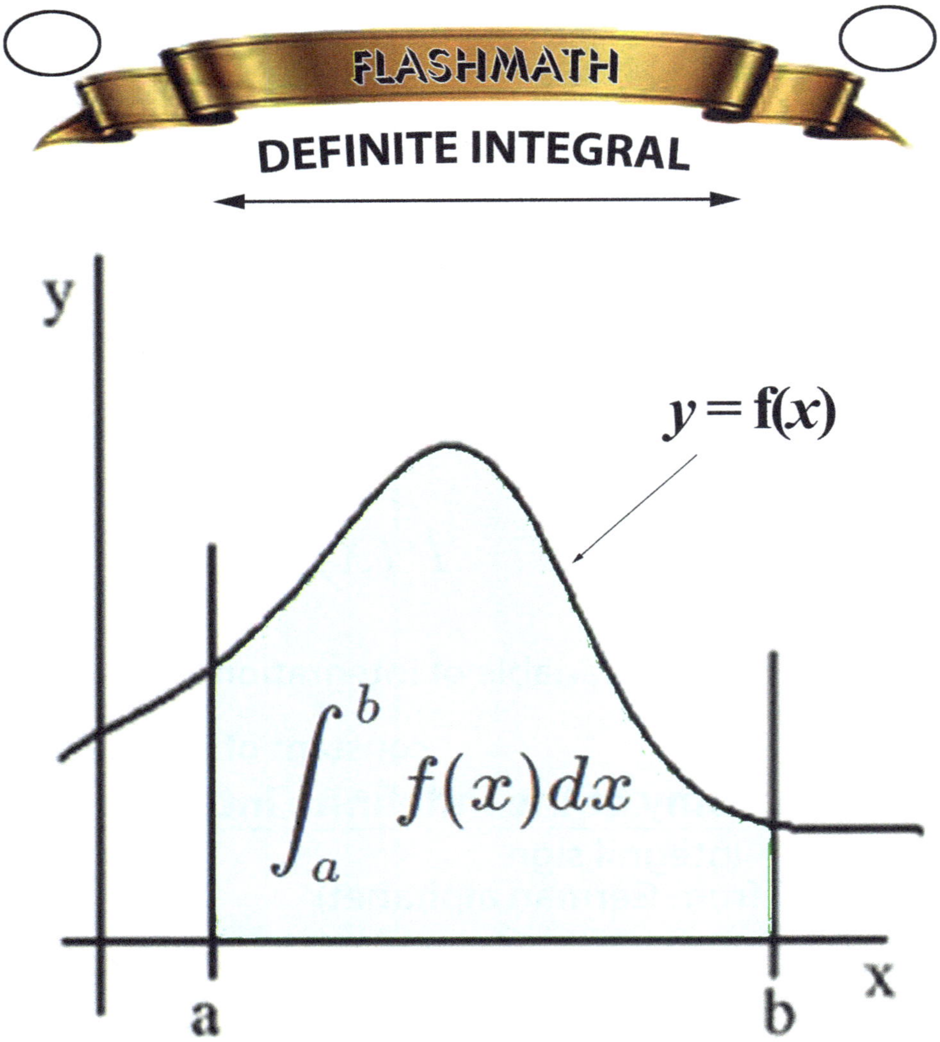

$$\int_a^b f(x)dx = F(b) - F(a)$$

Derivative of a function

FLASHMATH

DERIVATIVE OF A FUNCTION

The derivative of $q(x)$ is the rate of change of p with respect to x

$$q'(x) = \frac{dp(x)}{dx}$$

$q'(b)$ is the slope of the line tangent to $q(b)$ at $x = b$.

Let $y = f(x)$. The derivative of y with respect to x:

$$f'(x) = \frac{df(x)}{dx} = \frac{df}{dx} = D_x f(x) = \dot{y}$$

"prime" notation (Lagrange notation)

Newton's dot notation

Leibniz's notation

operator notation

Notation relating to derivatives

Laplace transforms

Laplace Operator (script L)

function of t $(t \geq 0)$

function of s

$$\mathcal{L}[\mathrm{f(t)}] = \int_0^\infty f(t)e^{-st}dt = F(s)$$

algebraic variable independent of t (treated as a constant in the integration)

$$\mathcal{L}[\mathrm{A}e^{at}] = A\int_0^\infty e^{at}\, e^{-st}\, dt = A\int_0^\infty e^{(a-s)t}\, dt$$¶

$$= \frac{Ae^{(a-s)t}}{a-s}\bigg|_{t=0}^{\infty} = \frac{A}{(a-s)}\left[e^{(a-s)\cdot\infty} - e^{(a-s)\cdot 0}\right]$$¶

$$= \frac{A}{(a-s)}[0-1] = \frac{A}{s-a}$$¶

Table of Laplace transform pairs

FLASHMATH

TABLE OF LAPLACE TRANSFORM PAIRS

$$\mathcal{L}[f(t)] = \int_0^\infty f(t)\epsilon^{-st}dt = F(s) \qquad t \geq 0$$

$$\mathcal{L}^{-1}[F(s)] = f(t) \qquad t \geq 0$$

TRANSFORM PAIRS

$f(t) \quad t \geq 0$	$\mathcal{L}[f(t)] = F(s)$
$A\epsilon^{at}$	$A/(s-a)$
A	A/s
At	A/s^2
At^n	$An!/s^{n+1}$ *
$A \cos \omega t$	$As/(s^2+\omega^2)$
$A \sin \omega t$	$A\omega/(s^2+\omega^2)$

*n = positive integer

$$\mathcal{L}[af(t) + bg(t)] = aF(s) + bG(s)$$

$$\mathcal{L}\left[\frac{df(t)}{dt}\right] = sF(s) - f(0)$$

$$\mathcal{L}[d^2f/dt^2] = s^2F(s) - sf(0) - df/dt\,|_0$$

$$\mathcal{L}\left[\int_0^t f(t)\,dt\right] = F(s)/s$$

$$\mathcal{L}[\epsilon^{at}f(t)] = F(s-a)$$

$$\mathcal{L}^{-1}[F(s-a)] = \epsilon^{at}f(t) \qquad t \geq 0$$

Tangent line, extrema, inflection

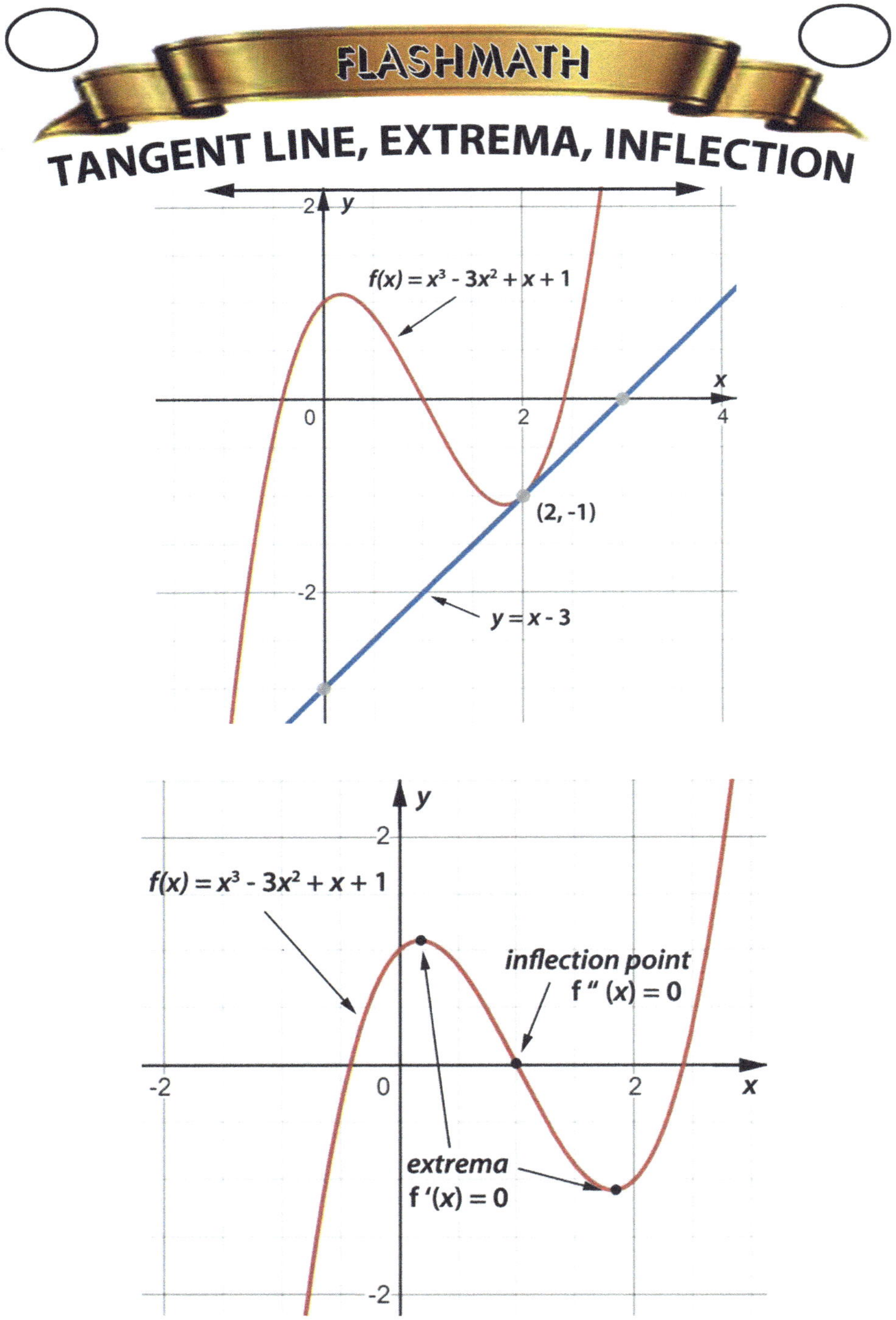

THE MATH TUTOR'S FRIEND

INFORMAL REVIEW OF HIGH-SCHOOL MATH TOPICS FOR TUTORS

- 9 -

' Series Guide '

Made in the USA
Monee, IL
02 November 2026

67385799R00066